THE UNDERGROUND ATLAS

The Underground Atlas

A gazetteer of the world's cave regions

by

JOHN MIDDLETON

& TONY WALTHAM

St. Martin's Press
New York

Library of Congress Cataloging-in-Publication Data

Middleton, John, 1921–
    The underground atlas.

    1. Caves.   I. Waltham, Tony.   II. Title.
GB601.M5   1987   551.4'47   87-16303
ISBN 0-312-01101-6

First published in Great Britain by Robert Hale Limited.

First U.S. Edition

10   9   8   7   6   5   4   3   2   1

Contents

Illustrations

Between pages 144 and 145

ACKNOWLEDGEMENTS

The authors would like to thank the following for permission to reproduce photographs: H. Fairlie Cuninghame: 1; S. Worthington: 8; S.C. des Causses: 10; I. Davinson: 16; J.M.H. Coward: 20; Akiyoshi-do Museum: 25; P.F. Wycherley: 26; A.J. Eavis: 27, 28, 38; Untamed River Expedition (D.W. Gill): 30; A.S. White: 31; C. Birkhead: 32; J.S. Corrin: 35; and A. Klimchouk: 39. All other photographs were taken by Tony Waltham.

Acknowledgements

The prime acknowledgement goes to Ernst Kastning of Radford University, Virginia, who produced most of the text concerning the United States.

The gazetteer originated from the files of the first author, accumulated when he was Foreign Secretary for the British Cave Research Association, and the authors are grateful to the BCRA Librarian, Roy Paulson, for his endless co-operation. Special thanks are also due to Alexander Klimchouk for his assistance with the text material on Russian caves, and last but not least to Claude Chabert for providing so much material and assistance.

THE UNDERGROUND ATLAS

Preface

Within this book, the world's known karst regions have been described under individual countries in an alphabetical sequence. Additionally some isolated areas are separated from their political host if they are significantly different or important. The continental maps in the introductory chapter shows at a glance how entries are to be found. For example, for Sardinia look under S, not I for Italy. Subdivisions within individual countries are normally listed so as to place the more important areas first where possible.

Local names are used as a general rule, except where an English translation is both widely accepted and better understood. Not all names can be placed on the maps, and the reader is referred to the Times Atlas for the locations of most regions, physical units and major towns. National caving organizations are mentioned when they are of particular significance.

Technical terms have been avoided except when their use is essential, and a short glossary is included for anyone not familiar with some of the more specialist caving terms. The cave maps, in either plan or profile, have all been been simplified from the original detailed surveys, and some abbreviations are used as standard: figures refer to depths in metres below, or above, the entrance level; entrances, sumps and major pitches (with their lengths in metres) are referred to by their initial letters.

Compilation of a volume of this nature has to involve a certain degree of subjective selection. It is, however, intended that all the major cave areas of the world have at least been mentioned, and it is hoped that their distinctive characteristics have been brought out in description. Only a limited number of caves have been individually featured; these are not necessarily the longest or deepest but are those considered by the authors to be the more important or more interesting.

It is hoped that this book will provide the reader with an insight to the wealth of caves and limestone regions around the world. By their nature, caves remain unknown until they are explored. Consequently the future will see the discovery of many more caves to add to the current statistics. And, tragically, some caves will be lost, when they are quarried away or perhaps drowned by new reservoirs. The loss of any cave is tragic because it is the loss of a precious and unreplaceable part of our environment. Caves provide a mass of scientific data for the geologist, they yield water supplies for people living in some limestone areas, and they provide sport for the growing numbers of cavers. They are a part of the natural world and as such deserve respect and reasonable treatment. That comes only from appreciation and under-standing – perhaps this book can promote some of that and thereby contribute to the conservation and future enjoyment of our underground environment.

The Continents

By way of introduction, the following paragraphs outline the broad distribution of limestone, karst and caves across the continents of the world. The accompanying maps also act as key and guide to the countries referred to in the main text. In general, political divisions have been used, but some isolated landmasses and large islands are separated out individually where the natural inclination would be to look under their own names and not the names of their parent countries. All the named countries featured in the main text appear on these maps, in their margins if isolated by sheer distance. Isolated islands which are not referred to have no known karst or caves.

EUROPE

The continent of Europe contains a karst of greater richness and diversity than perhaps any other area of similar size on Earth. All but four of the component countries can be considered to possess major karst regions, and within these are the world's deepest known cave in limestone, the longest and deepest in chalk, the deepest in conglomerate, the longest in sandstone, and others in lava and granite. The sport and the science of speleology both originated in Europe over one hundred years ago, and its people have always remained at the

forefront of developments. Few nations would consider that investigations into their karstlands have in any way reached maturity, and there is no doubt that Austria, France, Spain and Italy, at least, will yield many more systems of 1,000m or more in depth. In 1985 Europe contained nineteen of the twenty-two caves in the world explored to a depth of 1,000m; it also contained twelve of the world's twenty-six caves in which over 40km of passages have been mapped.

ASIA

The huge landmass of Asia embraces every type of terrain and natural environment known on Earth. Its limestone and karst are on a similarly dramatic scale, though the level of cave exploration is far behind that of neighbouring Europe, due to the constraints of politics, distance and resources available to be devoted to sport.

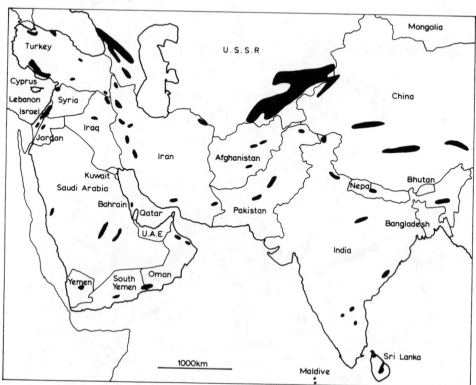

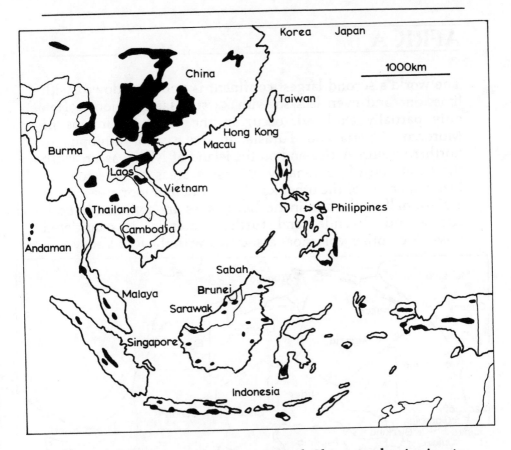

Enormous karst areas in Russia and China are beginning to reveal a wealth of caves which may be unmatched elsewhere in the world. Through the drier regions of south-west Asia, karst areas are isolated and scattered, with perhaps Turkey, Lebanon and Oman having the lion's share of the caves. The great mountains of the Himalayan ranges promise little more than disappointment for cave explorers, and the Indian peninsula is notably poor in limestone. South-east Asia, from China to the Indonesian islands, is the region of the future. Wild limestone mountains throughout the hot, wet tropical belt already have the largest – as opposed to longest or deepest – caves in the world. South-east Asia has the finest potential in the world for future cave discoveries, and the underground exploration is only just beginning.

AFRICA

The world's second largest continent is poorly endowed with limestone and even less so with karst. Major regions, often only partially explored, occur in the Atlas mountains of Morocco, Algeria and Tunisia, in the east of Ethiopia, in northern South Africa and on the island of Madagascar. With the exception of reasonably extensive areas of limestone in Namibia, most of the other karst is found in minor areas in the eastern half of the continent. Lava caves are common in both Kenya and Rwanda, and further potential is considered possible on other volcanoes associated with the Rift Valley.

Madeira
Canary
Morocco
Tunisia
Libya
Algeria
Egypt
Mauritania
Mali
Niger
Chad
Sudan
Djibouti
Senegal
Gambia
Guinea B.
Guinea
Burkina F.
Nigeria
Ethiopia
Sierra Leonne
Ivory C.
Togo
Benin
Liberia
Ghana
Bioko
Cameroon
Central A.R.
Equatorial Guinea
Sao Tome
Congo
Zaire
Uganda
Kenya
Somalia
Cape Verde
Gabon
Rwanda
Burundi
Seychelles
Tanzania
Zanzibar
Ascension
Angola
Malawi
Comoro
Zambia
Mozambique
St. Helena
Zimbabwe
Madagascar
Mauritius
Namibia
Botswana
Reunion
1000km
Swaziland
Lesotho
South Africa
Tristan da Cunha
Kerguelen

NORTH AND CENTRAL AMERICA

By combining the Caribbean region with North America, a continental unit is established to cover an enormous range of limestones, karsts and caves. The Canadian half of North America is, with some notable exceptions, poorly endowed with karst and caves. The United States has some large areas of karst, and though its caves nowhere achieve any great

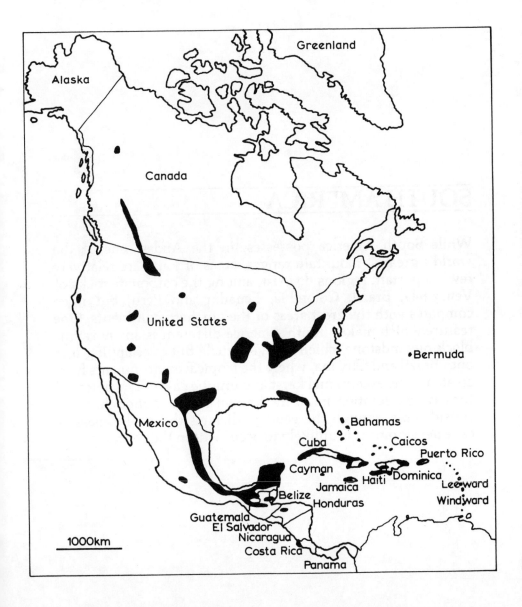

depth, they do dominate the world list of the longest known underground systems. In marked contrast, many of the Central American states and West Indies contain large proportions of limestone outcrops with some spectacular karst. Mexico, Belize, Cuba and Jamaica are major cave countries, and there is considerable exploration potential throughout the region.

SOUTH AMERICA

While South America possesses, in the Andes, one of the world's greatest mountain ranges, areas of karst are relatively few. Important regions do exist, among the carbonate rocks of Venezuela, Brazil, Colombia, Ecuador and Peru, but none compares with the great areas of the northern continents. One feature which makes South America different is an enormous block of sandstone, chiefly in Venezuela but overlapping into both Brazil and Guyana, where the tropical environments have created some exceptional karst and unique caves. The potential for cave exploration in all the known karsts is thought to be considerable, though the geology dictates that really massive cave development is unlikely to occur within the continent.

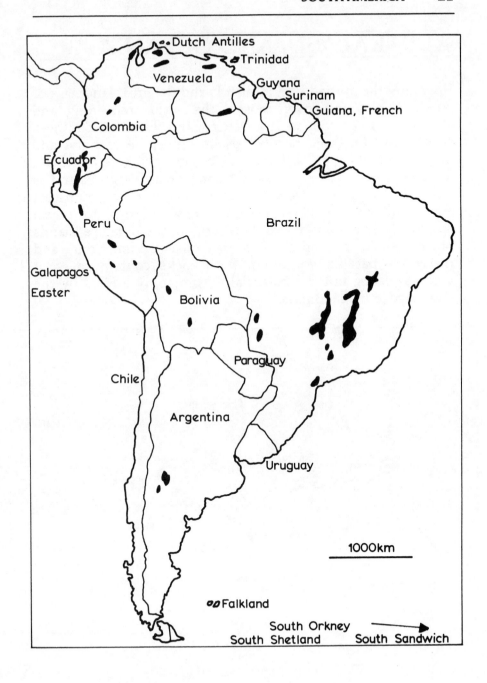

Dutch Antilles
Trinidad
Venezuela
Guyana
Surinam
Guiana, French
Colombia
Ecuador
Peru
Galapagos
Easter
Brazil
Bolivia
Paraguay
Chile
Argentina
Uruguay
1000km
Falkland
South Orkney
South Shetland
South Sandwich

AUSTRALASIA AND OCEANIA

Spanning the myriad Pacific islands and the great landmass of Australia, this unit covers most of the world's remaining land except for ice-shrouded Antarctica. Like the other southern continents, Australia hardly has its fair share of karst and caves, though the Nullarbor is a glorious exception and Tasmania has proved a major caving area. New Zealand is richer in caves, and New Guinea relates more to south-east Asia with a vast karst potential now starting to reveal spectacularly massive caves. Further east, the paradise islands of the Pacific do contain scattered areas of limestone, and many are basaltic volcanoes. Both rocks contain caves at various places, but the potential is commonly limited by the small size of the landmass.

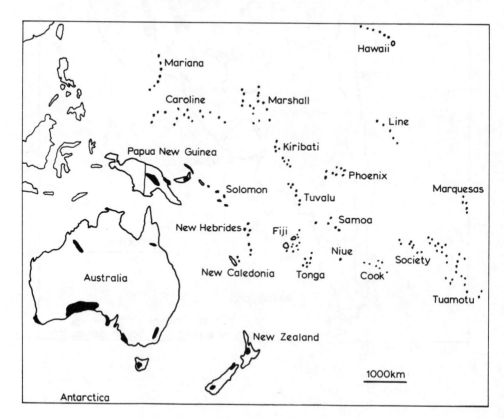

The Countries

AFGHANISTAN

As befits a nation sitting astride the join of the Iranian plateau and the Himalayan ranges, Afghanistan has a landscape of desolation and grandeur. Karst rocks are responsible for little of this design, and limestone is known only on the high tablelands to the north of the Bamiyan valley, on the massif of Salang (Parwan) and on the lower desert massif of Bolan (Zabul). Current knowledge is limited, and extensive cave development does not seem likely. The longest system is the Ab Bar Amada (Parwan) with 1,220m of passage explored by a French team.

ALASKA

While a considerable proportion of the USA's forty-ninth state is underlain by limestone, few karst features have been noted and all the known caves are small. The many extensive glaciers hold, perhaps, greater potential for the glacio-speleologist, although the Boundary Range of the Juneau Icefield is the only region to have received any serious investigation. Here Lake Linda Cave on the Lemon Glacier proves the longest at 500m.

ALBANIA

The small Balkan republic of Albania, situated within the great Dinaric mountains and bordered by the eastern Adriatic Sea, contains some of the least known and wildest country in Europe. Some two-thirds of the land rises above 800m. Limestone predominates, particularly in the northern half, where tilted massifs, cut by deep glaciated valleys, terminate in jagged alpine peaks. Major karst features are found between the central Deja mountains and the Prokletije range which forms the country's northern boundary. Caves are known but specific details are at present unobtainable. Access is severely restricted and there are no organizations specializing in any aspect of karst study.

ALGERIA

The Atlas mountains of northern Algeria contain important and extensive karst areas in north Africa, second only to Morocco. At the western end of the range the Monts de Tlemcen have many depressions, sinks and springs, and also the country's longest cave. The Rhar Bou Maza, over 14km long, takes the Tafna river underground through magnificent large river passages; unfortunately much of the length lies beyond a 60m sump. Karst in the spectacular Ouarsenis massif is rather less developed and includes only small caves.

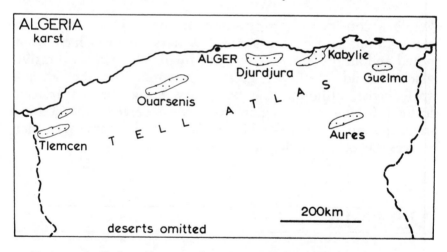

Finest of all the Algerian karst is the Djurdjura massif, with a rolling terrain of dolines and lapies, and numerous caves and potholes. The deepest system is Anou Ifflis, with a single streamway descending steeply to a sump at −975m, recently passed, to more shafts reaching depths below −1007m. In the same karst lie spectacular Anou Boussouil and Anou Timedouine, 476m deep with a 190m shaft in it. The lower mountains of the Petite Kabylie have similar karst landscapes and also many sea caves. Further east the Guelma area is again karstic and includes the fine Ghar el Djemaa, 110m long and

120m deep. South of the Tell Atlas, the Aures massif is a karst, but its only known caves are on a small scale.

Cave exploration in Algeria has been largely by French expeditions, though there have been several local clubs.

ANOU BOUSSOUIL

Algeria's second deepest and perhaps most notable system is situated in the scenic massif of Djurdjura. Its large entrance lies some 6km west of Tizi-N'Kouil village and just 200m from route N33. Entry was first made by local cavers in 1933, but the presumed terminal syphon at −505m was not reached until 1947 after visits by French cavers. The system consists of a clean series of flood-prone shafts and cascades, each terminated by deep pools. In 1982 an expedition from Toulouse extended an upper gallery down more pitches to a new depth of −805m and combined length of 3km. This new section includes a chamber, the Salle des Bermudes, 380m in circumference and 70m in height. The water is thought to resurge 8km to the east, in the valley of Aït Ouarbare.

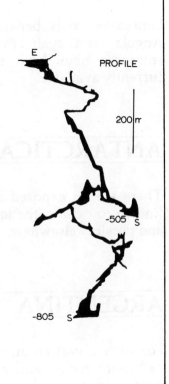

PROFILE

200 m

−505 S

−805 S

ANDAMAN AND NICOBAR ISLANDS

Surface karst features are known on many of the coral islands composing this Indian Ocean group, but no caves have so far been recorded.

ANDORRA

Of no speleological interest

ANGOLA

Limestone exists beneath the dense bush of the centre of Angola, and again in the south, where large river-cave entrances have been noted. Little further information is currently available.

ANTARCTICA

There are no exposed carbonate rocks but glacier caves are known to exist in the ice of the Trans Antarctic Mountains – and possibly elsewhere.

ARGENTINA

For such a vast country, Argentina is poorly endowed with carbonate rocks, and of the regions which do exist, none contains particularly well-developed karst. Many small caves can be found in the north-west of La Pampa province and in southern Mendoza (Cueva de la Bruja, 2,500m long), whilst in northern Neuquen, the Cueva del Leon (800m) occurs in a small area of gypsum. The long volcanic chain of the Andes has been little investigated for caves; currently the longest lava cave is the Cueva de Doña Otilia (Mendoza, 838m), but it does seem inevitable that more such systems remain to be found.

ASCENSION ISLAND

Of no speleological interest.

AUSTRALIA

This fascinating continent, so rich in exciting physical features, has rather limited areas of both carbonate rocks and karst. Over 4,500 caves are currently known but, while many are exceptionally well decorated, present unusual exploration problems or simply provide exhilarating caving, none is of major proportions on a world scale. Local cavers have been active since the 1940s.

Six regions effectively cover both the diversity of Australian speleology and the majority of its caves. Of these, Tasmania is

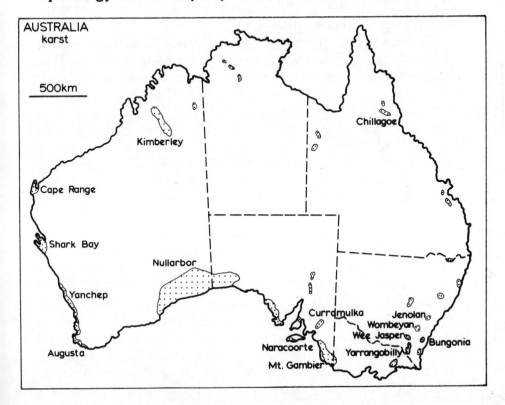

AUSTRALIA
karst

500km

Kimberley

Cape Range

Shark Bay

Nullarbor

Yanchep

Chillagoe

Currdmulka

Jenolan

Wombeyan

Wee Jasper

Bungonia

Naracoorte

Yarrangobilly

Augusta

Mt. Gambier

undoubtedly the most important, but perhaps even better known is the Nullarbor Plain. This is one of the world's largest continuous areas of limestone, covering over 200,000 sq km. It stretches from the Great Australian Bight, in the south, inland for over 300km to the Great Victoria Desert. The semi-arid terrain is monotonously flat with almost no obvious features, which makes exploration and cave identification doubly difficult. There are neither surface stream or dry valleys, and the only karst consists of isolated dolines. As compensation for its bleak surface, the Nullarbor possesses some unique caves among the two hundred or so which have been explored to date. Most of the cave entrances are situated at the bottom of large dolines from which massive tunnels penetrate horizontally. These then enter equally large chambers before great lengths of flooded passage present divers with major logistical challenges. Many systems reach around 1km in length, but the two classic caves are the single 6km-long passage of Cocklebiddy Cave, Naretha, of which 5,040m is submerged in three long sumps, and Mullamullang Cave, Madura (11km long, 125m deep), containing an enormous tunnel over 4km in length.

Chillagoe, in northern Queensland, lends its name to a band of limestone roughly 160km in length by 8 km in width. This has been sculptured into many remarkable towers which reach 150m high and over 1km long. More than 370 caves and potholes have been explored in these bluffs, and many have been only partially investigated. The major system, within Queensland Bluff, is the Queenslander-Cathedral Cave, a complex of several linked caves totalling over 6km. A problem of exploration within these caves is the internal temperature of 30°C, but the challenging and varied caving, the geology of the surroundings, and the large populations of swiftlets and bats make this a karst of exceptional interest.

The wild and colourful Kimberley Ranges, in the north of Western Australia, are potentially Australia's greatest cave region, but their distant isolation has, to date, prevented many studies. Limestone covers an area of roughly 8,000 sq km, which includes a spectacular and well-developed tropical karst of towers, small poljes, dolines, lapies and major underground features. Most of the known caves are predominantly horizontal, and Mimbi Cave extends for over 8,500m, although 2,000m of this has no roof, being at the bottom of a deep rift in an extensive field of lapies.

The south-eastern cave region is a composite of the many small limestone outcrops among the mainly igneous rocks of eastern New South Wales. Included within this are such well-known areas as Jenolan, Bungonia, Wombeyan, Yarrongobilly, Wee Jasper and many others. Few exceed 20 sq km in extent, and the majority are much less, but their location close to several cities has ensured active exploration. Wombeyan is one of the smaller karsts, but some 240 caves have been recorded, many of exceptional beauty but none of great extent. Bungonia, with its impressive 400m-deep gorge, can lay claim to the majority of the mainland's deep caves, although Odyssy Cave, the deepest, reaches only 148m. The deepest cave, Eagles Nest System (−174m and 3,600m long), lies in Yarrongobilly, but apart from the 139m-deep East Creek Cave there are few others of note nearby. The sandstone hills of Braidwood contain the spectacular 30m-diameter shaft of Big Hole, which descends to 113m, while in the Kanangara-Boyd National Park limestone, the Lanigans-Onslow Cave is over 6,000m long. The decorated caves of Jenolan had the distinction of becoming Australia's first cave reserve in 1860, and the show cave of Marghevila in 1880 became the world's first to use electric lighting.

The last major cave region is the long coastal strip of karst, notably around Yanchep and Augusta, in Western Australia. Many horizontal caves have been found and almost all are noted for their stunning formations. The 1,900m-long Jewel Cave contains straws up to 5.8m in length whilst Easter Cave (7,000m long) is an exceptionally well-decorated water-table maze. Broken limestone outcrops extend up the coast as far as Shark Bay and the Cape Range, but few caves are known.

In the remaining karsts there are few caves of significant size. South Australia has the best, with the fossil caves of Naracoorte, Corra-Lynn Cave over 5km long at Curramulka, and the great flooded potholes and cenotes of Mount Gambier.

Tasmania

With over sixty areas of known karst, and the possibility of others lying beneath the dense and often little-explored rain-forest, Tasmania is clearly Australia's premier caving state. It contains most of the nation's deep caves, and while none is of major proportions, some rank among the world's classics.

Three major cave areas are currently known, all fairly accessible and not far from major cities. West of Hobart, the Junee Ridge and Florentine Valley karst contains most of the island's deepest, wettest and most sporting caves, with seven reaching 200m in depth. The Ice Tube entrance to the Growling Swallet stream cave is now the deepest and is just one of a group of fine shaft systems. A little to the south, Khazad-Dum is a magnificent stream cave with twelve waterfall pitches on the way to a depth of 321m; the shaft system of Darrowdelf connects into its final chamber to offer a fine through trip, but the water resurges 4km away at Junee.

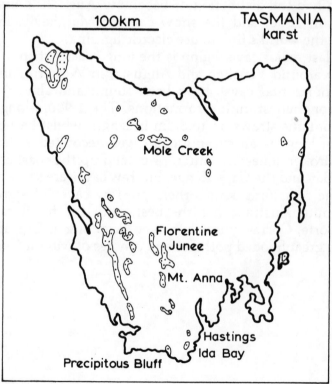

100km   TASMANIA karst

Mole Creek
Florentine
Junee
Mt. Anna
Hastings
Ida Bay
Precipitous Bluff

Close to the south coast, the Ida Bay karst is not so extensive but does contain some fine caves. The Exit Cave complex is 17km long with some massive passages and is also noted for some long gypsum needles and its own glow-worm population. Mini Martin is a 170m broken shaft which drops into the main Exit streamway. Mystery Greek Cave carries the main sinking stream and is not connected to Exit, though it does have its own shaft entrance through Midnight Hole.

GROWLING SWALLET SYSTEM

Situated to the west of Florentine Peak, in the Junee Florentine karst, this complex stream cave system is currently Australia's second deepest, at −345m. Its horizontal extent is well in excess of 7km, and it is possible to make various traverses between the different entrances, although in practice exit is usually made through Growling Swallet as this cave remains partially rigged. The classic trip, and arguably the finest in Australia, is that made from Ice Tube to Growling. This covers almost 4km of passage and descends the full 345m depth. Ten airy but very wet pitches have to be negotiated before a very awkward meander connects to the main streamway. There then follows a series of tricky climbs, crawls and narrow passages, before 130m of ascent reaches the main Growling sinkhole.

A number of tributaries join the main stream, and two now provide entry routes. Slaughterhouse Pot is initially tight but opens up onto five short pitches separated by a complex boulder choke. This connects with Growling down an aven at −230m. Pendant Pot is the most recently discovered entrance and involves a quick descent to a sump at −192m. Divers can then make a connection through a 15m-long and 5m-deep syphon into the main cave.

PLAN

500m

The Mole Creek karst, in the north of Tasmania, has over two hundred caves, many noted for their excellent decorations. Kubla Khan is the most spectacular, with 2,500m of passages and chambers containing superb calcite formations. Nearby, Herbert's Pot is over 5km long.

Of the remaining karst areas, many have been little investigated. The Mount Anna karst has 600km of relief in dolomitic limestone and has recently yielded Anne-a-Kananda; this is a complex cave leading down to some deep shafts which achieve a depth of 373m, currently the Australian record. The Hastings dolomite karst also holds promise, with Wolf Cave already mapped for over 2km, and Precipitous Bluff is a potentially exciting karst with some spectacular river sinks, which, however, need a two-day approach walk.

AUSTRIA

In a country where over two-thirds of its area is classed as alpine and almost half of that is composed of various limestones, it is not surprising that Austria ranks as one of the world's greatest caving regions. Karst rocks occur mostly in the northern Alps, which straddle the spectacular Salzburg province.

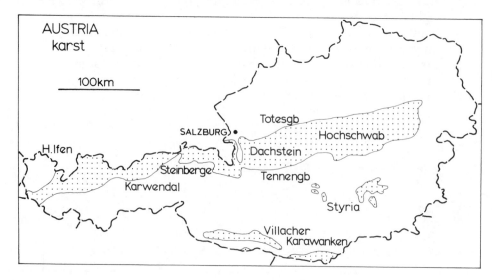

Austrian caving history can be traced as far back as 1387 and can boast the world's first speleological association (Vienna, 1879). In 1922 a Speleological Institute was founded, and in 1949 a national body was organized. There are now numerous regional groups. Visitors should first contact the national organization for information on local groups. Apart from obvious safeguards for sporting trips, serious prospectors have to abide by certain extra rules.

The Northern Alps

These mountains extend from the western edge of Vorarlberg, 450km eastwards to the Vienna basin. In the central area they reach up to 70km in width. The morphology varies considerably, and there are exceptions to the regional patterns. In the west (Allgauer, Lechtaler, Karwendel etc) steeply dipping beds have created precipitous mountains with jagged peaks and only a small extent of obvious karst. Throughout the centre (Salzburg and western Steiermark) large, gently dipping blocks of massive limestone have been dissected by glacially scoured valleys. This has resulted in extensive plateaux (Tennengebirge, Dachstein, Totesgebirge etc) each deeply etched by lapies, dolines, shafts and sinks, while caves and potholes form complex mazes within. Similar characteristics, though slightly less pronounced, extend eastwards to the limit of the limestone. Along much of the northern edge of the southern Alps run the pre-Alps. These consist of heavily forested hills which can reach up to 1,500m, and small caves, dolines and springs are common.

Excluding the caves within the province of Salzburg, most other great systems are to be found close to its boundaries. In Upper Austria, the 400 sq km Dachstein massif averages 1,800m in height. The Dachstein limestone reaches around 1,000m in thickness and contains numerous caves. The principal system is the ever-lengthening Dachsteinmammuthöhle, a vast maze containing many large galleries now over 35km long; it reaches to −727m and +423m, giving a total depth of 1,170m. Close by is the Dachsteineishöhle, a 2,000m-long show cave containing some of Europe's most spectacular ice formations, and also the Hirlatchöhle, a 38km-long system at low-level. The Hollengebirge, emerging from the Upper Austrian pre-Alps, also has massive, gently dipping bedding. The outstanding pothole of this mountain is the Hochleckengrosshöhle (−896m, 5km long) containing the unbroken 351m-deep Steierwascheurschacht. On the border between Upper Austria and Steiermark lies the great 590 sq km Totesgebirge karst massif. Both long and deep caves are to be found in profusion, dominated by the complex Raucher-karhöhlensystem (742m deep, 38.7km long), the three-entrance Feuertalsystem (−913m, 6,200m long), the short, sharp Trunkenboldschacht, with its 550m of almost continu-ous pitches to −859, and the Stellerwegsystem (−972m).

Further east the most important speleological region is the Hochschwab. Here, beneath a plateau terrain of dolines and depressions interspersed with alpine meadows and dwarf pines, lie the large passages of the six-entrance Frauenmauer-Langstein Höhlensystem (−580m, 19,340m long). This area and the nearby Schneebirge massif have received much study, as they are a major water catchment for Vienna.

West of Salzburg, areas of extensive karst are uncommon. The Karwendalgebirge is noted for its high-level (2,000m+) lapies and dolines. There are also very large resurgences around its base. On the border between Vorarlberg and West Germany is the Hoher Ifen massif, renowned for an exceptionally fine 10 sq km of deep lapies and small caves.

The Salzburg Province

The northern half of this Austrian province currently contains a concentration of major systems greater than perhaps any other area of comparable size in the world. At the start of 1984 there were forty known caves exceeding 1km in length and thirty over 300m in depth. Almost all of these occur in the massifs adjacent to the West German border, with the exception of the more central Tennengebirge. This massive tabular block rears up from the surrounding valleys with fearsome verticality, and its barren surface is pitted with deep shafts, dolines, caves, depressions, lapies and dry valleys. Major new discoveries are made each year with amazing

SCHWERSYSTEM

The 1,219m-deep cave is located on the slopes of the Schwer valley in the Tennengebirge at an altitude of 1,866m. It consists of a progression of short pitches leading to a horizontal section and more pitches before the 205m-deep Puits Boeing. A final tortuous passage leads to the bottom. The cave is both technically difficult (requiring around 1,400m of rope) and physically arduous, due to constrictions, traverses and the extremely cold conditions. The water resurges at the Winnerfall.

Exploration was commenced in 1979 by the Spéléo Club de Marseilles and a depth of 254mm reached. Further visits over the following three years saw exploration reach the current depth and a length of 4,065m.

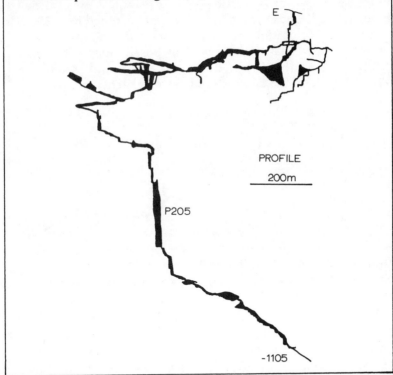

E

PROFILE

200m

P205

-1105

regularity, and some of its great caves are among the world's deepest, notably Schwersystem, (1,210m deep). Schneeloch reaches levels of +132m, −969m, with its single evenly descending streamway; the Berger-Platteneckhöhlensystem is a complex covering a vertical range of 937m, and the renowned ice cave of the Eisriesenwelt is 42,000m long,

LAMPRECHTSOFEN

This great Austrian system stands out from others because it has been totally explored from its resurgence upwards for an unparalleled distance. Its entrance lies in the Saalach valley at an altitude of 664m, on the slopes of the Leoganger Steinberge. A short 'tourist series' (i.e. show cave) leads to an extremely wet section of the cave which is highly prone to flooding. Once past, persistent explorers have ascended great shafts, chimneyed up fissures, dug through boulder chokes, negotiated mazeways, explored some 14,650m of passage and reached a point 995m above the entrance (1,005m in total). The cave was first investigated in 1833, and in 1905 was developed for tourism, 1965 saw the first serious progression upwards, and by 1969 a height of +511m had been reached. The current statistics were recorded in 1980.

Surface exploration in an attempt to find a second and upper entrance has yielded several possibilities. The most promising of these are the 730m-deep and 4km-long Wieserloch, and the 560m-deep Lofererschacht.

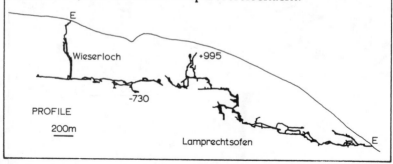

though ice occurs only in the entrance series.

Most of the major karst regions ranged along the Bavarian frontier are of a character similar to the Tennengebirge, and all are in the early stages of exploration. In the Loferer Steinbirge there is the Herbsthöhle (−684m), in the Leoganger Steinbirge, the ascending Lamprechtsofen and, in the Untersbirge, the Salzburger Schacht (−606m). The Hoher Göll has the Gruberhornhöhle (894m deep) and Jubiläumsschacht, and the Hagengebirge contains the Jägerbrunntrogsystem and the nearby Tantalhöhle (30,850m long). Lesser systems exist in the remaining massifs, but again exploration is in early stages. A new discovery in the Steinernes Meer is the Kolkbläser Monsterhöhle System (over 11km long, 532m deep).

JUBILÄUMSSCHACHT

This recently explored Austrian system is to be found on the steep slopes of the great cirque of the Gruberhorn in the massif of the Hoher Göll. Its entrance (altitude 2,009m) is situated in a deep fissure in the sharply eroded lapies. A short, steep descent leads to a room containing snow before a larger horizontal passage enters a large chamber. From here a series of great (70m, 74m, 128m, 126m, 201m) and small pitches rapidly descends to a depth of 900m. A twisting and often tortuous canyon more than doubles the length of the cave and, after several short pitches, reaches the terminal syphon at −1,173m, the total length being 2,380m.

The cave was discovered and descended to a depth of −459m by a team of Polish cavers in 1980. An enlarged team reached the bottom the following year with one bivouac at −900m. There are reported to be several unfinished ends to be fully explored.

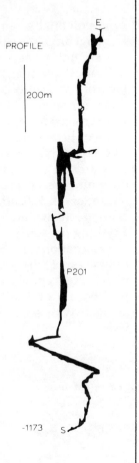

PROFILE

200m

P201

-1173   S

JÄGERBRUNNTROGSYSTEM

This potentially remarkable system is situated high on the Hagengebirge massif in the Salzburg province of Austria. It has six entrances – the upper and lower Petrefaktencanon (altitude 2,138m and 2,132m), Zwillingschacht (alt. 2,066m), Jägerbrunntroghöhle (alt. 1,884m), Ochsenkarschacht (1,863m) and Sulzenkarfischöhle (alt. 1,828m). All but the Ochsenkarschacht are found relatively close together, and all connect before descending to the present depth of 1,078m. The complex of ever-lengthening passages, which extend outwards in three directions, currently covers 30,810m. At one point the cave is only 400m distant from the Tantalhöhle (entrance alt. 1,710m), itself 30,850m long and 435m deep. A connection would make the system 20km longer than anything presently known in Austria.

Exploration of the cave commenced in 1955 with the discovery of the Ochsenkarschacht, but it was not until Petrefaktencanon was found in 1976 that the full possibilities of the system were appreciated, and this was pushed, with some very hard caving, over the succeeding years.

PLAN

500m

The Central and Southern Karsts

The central karst of Styria lies to the north and west of Graz for a distance of up to 45km in a region of flat-topped, wooded hills, up to 1,400m in height. Many of these terminate in abrupt cliffs. Dolines occur around the 700m level and there are numerous caves. Notable amongst these are the 5,975m-long Lurhöhle show cave, the Katerloch (1,500m long, partly show cave) and the enormous passages of the Drachenhöhle. Many caves are rich in palaeontological material.

The southern karst forms two areas of interest in the Villach Alps and in the larger Karawanken range, both in southern Carinthia. The former is the most important and, in spite of its being thickly wooded, there are many small caves, large dolines and impressive field of lapies. The Karawanken overlaps into Yugoslavia, and it is on that side that karstic features really predominate.

AZORES

A volcanic archipelago containing several lava caves – the longest (2,700m) being the Gruta dos Balcoes, on the island of Terceira. On Pico there are more short but notable caves in basalt lava.

BAHAMAS

Extending for over 800km, this chain of coral islands is renowned for its interesting and often unusual erosional features. Unfortunately, due to the islands' low relief (rarely greater than 3m), cave development is dominantly below both the water-table and sea-level – in the famous Blue Holes, named from the colour of their deep water surrounded by the shallow coral flats. The major systems already explored occur on Grand Bahamas – Lucayan Caverns by Freeport (9,180m long), Zodiac Caverns and associated systems by Sweetings Cove, and others around Andros Island. Cave entrances have been noted, but not yet explored, at depths of over 100m.

BAHRAIN

Limestone is the dominant rock of the islands but karst features are minimal and no caves are known.

BALEARIC ISLES

All these Spanish islands are composed chiefly of limestones, most of which support karst to some degree. Lapies is extensive throughout, while on the two larger islands, poljes, depressions, shafts and open caves also occur. The longest caves are on Majorca, in the hills between Espotas and Escorca in the west and around Aita in the east. The deepest (−317m) is the Cova de sa Campana, Escora, and the longest (1,700m), which has also been developed as a spectacularly well-decorated show cave, is Cova del Drach, Arta. Many caves have yielded important archaeological finds.

BANGLADESH

Of no speleological interest.

BELGIUM

The rolling hills of the south-east contain over two thousand known caves. Most of these are situated to the south of Namur (Trou Bernard, −140m), south of Rochefort (Grotte de Han, 5,720m), south of Liège (Grotte de Remouchamps, 2,800m) and by Hotton (Grotte de Hotton, 3,500m). Surface karst consists principally of verdant dolines, depressions, dry valleys, sinks and resurgences. Potential for new exploration is small, but occasional finds are still made. The upstream part of the river passage of the Lesse at Han, was found only in 1972 (Gouffre de Belvaux, 2,560m long).

Local cavers are extremely active, both nationally and internationally.

BELIZE

While this country is small, it does, like its neighbour Guatemala, contain a karst which is both extensive and diverse. The south is dominated by the granite Montañas Mayas. This in turn is surrounded to the east, north and north-west by a broken limestone platform upon which a rich tropical karst has developed. Cones, towers, depressions, poljes and active river caves occur in profusion, and particularly notable among the caves is the very beautiful Actun Lubul Ha (almost 4km long and 113m deep). It is situated, along with many similar systems, south of San Antonio. The Mountain Pine Ridge region of the north is of particular interest due to the many small areas of limestone which rest directly on the granite. Here, great and mostly unexplored rivers have cut completely through the hills, and slightly further west, in the Chiquibul river area, more than 24km of cave passages have recently been discovered. These include 11km in Actun Tunkul, a cave notable for the fact that most of its passages are 40m wide; it also contains Belize Chamber, 400m long, up to 200m wide and among the largest in the world.

The second major district is that which extends between the River Sibun and the sea. This is a lowland karst of little-investigated cones and towers. Several short, beautiful caves are known, many of which have yielded important archaeological discoveries. Vegetation and insects made the region very difficult to explore.

Of the remaining carbonate lands, the north's are similar to those at Yucatan in Mexico, with less-developed conical hills, shallow depressions, occasional dolines and caves. The many low-lying coral offshore islands all show karst development to some degree and the outermost of these, Lighthouse Reef Lagoon, contains one of the first explored sea Blue Holes (some 140m in diameter, 110m deep), draped with many massive formations.

Exploration in this exciting country, whose potential has only recently being appreciated, has been chiefly by expeditions from the USA and Canada.

BENIN

Of no speleological interest.

BERMUDA

This group of low-lying coral islands contains more than a hundred known caves, of which the largest, the Green Bay Cave System (1,340m), is situated by Hamilton on Great Bermuda. Many of these caves are accessible only to divers.

BHUTAN

There are no known areas of karst.

BIOKO (Fernando Póo)

Volcanic islands in the African Bight of no speleological interest.

BOLIVIA

The mineral-rich mountains of this country contain at least two areas of little-investigated limestone. Neither of these is thought to contain any great potential but their inaccessibility currently precludes detailed exploration. The area around the village of Torotoro in Potosi contains the country's longest and deepest cave, the Gruta de Umajalanta (−130m, 1,620m long); the other region, which has yet to yield any caves of importance, is in La Paz near Sorata. A further small area in the Nevada de Chocaltaya is notable for the great altitude of its caves, such as the Cueva de Chocaltaya at an elevation of 5,400m.

BOTSWANA

A small band of dolomitic limestone extends from the Transvaal between the towns of Gaberones and Lobatsi. It is rolling hill country rich in lapies, sinks and small caves. In the Aha and Cwihab hills close to the Namibian border are several large caves, such as Independence (−70m) and Drotsky's (1,200m long).

BRAZIL

This great country, which occupies almost one half of all South America, contains an area of limestone karst which comfortably exceeds all those found within the remainder of the continent. Over a thousand caves have so far been explored by enthusiastic and rapidly expanding groups of local speleologists. Of these caves, some fifty are known to extend for over 1km, and sixteen reach 100m or more in depth. Further potential for horizontal development is thought considerable, but depths greater than 300m are unlikely ever to be reached. Most caves occur in limestone but a significant minority are found in sandstone (with more possibilities in Roraima), schist, gneiss, granite and basalt. Access to all the caving regions is generally difficult due to dense forest, and most discoveries are situated close to roads and habitations.

The principal region, with more than two hundred caves, is in the states of São Paulo and Parana. Speleologically the area is known as the Vale do Ribeira; it extends from the town of Curitiba to Avare and includes all the Serra Paranapiacaba. This is where almost all the country's deep caves are situated, particularly in the vicinity of Iporanga and Apiai (Abismo do

Juvenal, −252m; Caverna do Ribeiraozinho, −220m).
Reasonably long systems are also known, including the
5,680m Caverna do Santana.

Two large and three smaller areas of karst, speleologically
grouped as the Bambui region, are found within the states of
Bahia, Minas Gerais and Guias. This is the home of the
country's longest systems and includes the Conjunto São
Mateus Imbira, São Domingo, Goias (20,540m) and, in the
same area, Conjunto Angelica-Bezerra (9,775m). In the
municipality of Corumba de Goias, the Gruta dos Ecos cave
(1,500m long, 115m deep) has developed in a mixture of
micaschist and quartzite.

In all the other karst regions caves are known but, with the
exception of the 1,300m-long Gruta Racardo Franco (Mato
Grosso South), none extends beyond the 1km length or 100m
depth. Surface karst generally is rarely spectacular, although
large dolines have been noted in Mato Grosso South.

Roraima

Bambui

Mato
Grosso S.

BRAZIL
karst

Vale do Ribeira

500km

BRUNEI

There are no known areas of karst.

BULGARIA

Over three thousand known caves are distributed throughout the country's three major karst regions. Most of these systems are of relatively minor proportions, but the potential does remain for considerable extensions. Over half of the caves occur within the great central belt of mountains known as the Stara Planina, and more specifically in the district around Lovec (Raychova, −382m), Vratsa (Temnata Doupka, 7,000m long) and Sliven (Yamato na Kipilova, −350m; Lednika, −242m). To the north, the rolling plains of the Danube valley contain a further quarter of the known caves, with the area south of Ruse being the most important (Orlova Tchuka, 13,150m long). The deepest known cave in the country is now Barkite 14 (415m deep).

The much wilder and more rugged Pirin and Rodopi Planinas of the south have most of the remaining caves and perhaps the greatest potential for both depth and length. The little-explored region around Smolyan already has one known cave of 6,450m (Mamova Doupka), and the 192m-deep Kazana, in the northern Pirin Planina, is in an area with a depth potential in excess of 500m. Of the several small, isolated areas, the most important is that not far from Pernik, where the country's longest cave is situated, the 15km-long Duchlata.

Speleology is very popular within the country, and there are many local groups.

BURKINA FASO

Of no speleological interest.

BURMA

Very large areas of limestone exist within the country, but little is positively known of its karst. Much of the vast, deeply dissected plateau of Shan is limestone, and in some areas both tower and cone karst have been reported. References to caves can be found for the central district of the northern state of Achin, and isolated cones of limestone similar to those in neighbouring Thailand are to be found on the long peninsular state of Tenasserim. It is reasonable to assume that similar features exist in the eastern border states of Kayah and Kawthoolei.

BURUNDI

There are no known areas of karst.

CAICOS ISLANDS

The extensive limestones of these West Indian islands are rich in both caves and karst, and while the caves are generally short, Conch Bar Cave extends for over 2,000m.

CAMBODIA

Carbonate rocks have been noted beneath the densely forested south-western Cardamones hills. No karst research has been recorded and no caves are known.

AUSTRALIA Weebubbie Cave, Nullarbor

AUSTRIA: Dachstein Eishöhle

BELGIUM: Grotte de Père Nöel

CANADA:
Castleguard Cave

CANADA:
Castleguard Cave

CHINA:
Nanxu river sink and arch,
Guangxi

CHINA:
Guanyan, Guangxi

ETHIOPIA:
Sof Omar

FRANCE:
Gouffre Berger

FRANCE: Grotte Amélineau, Causses

FRANCE: Grotte Gournier, Vercors

FRANCE: Mas d'Azil, Pyrenees

GREAT BRITAIN: Shatter Cave, Mendip

CAMEROUN

No areas of karst are known but deep tectonic fissures have been recorded in the mountains of the west. It is also possible that lava caves may exist on the larger volcanoes of the same ranges.

CANADA

The vast land area of Canada contains only a small proportion of karst. Limestone and gypsum do exist; in much of central Canada the lack of relief hinders significant cave development, and in many parts karst processes are further restricted by the climate. Cave exploration is also limited due to the difficulties of access posed by sheer distances. However, there are isolated areas of karst and caves which, particularly in the southern Rockies, are of major importance.

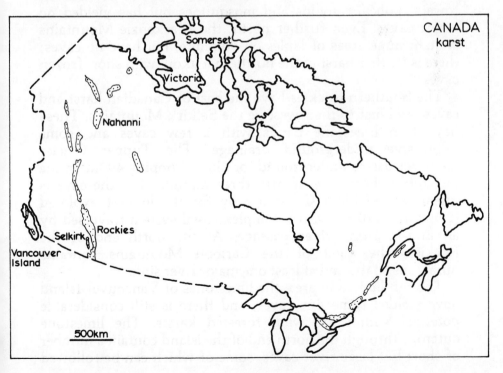

The Atlantic provinces have isolated gypsum and limestone, with a few short caves, but the potential is very limited. The geology is similar in Quebec, though the longest cave there, in the south of the province, is the Grotte de Saint-Casimir with a through streamway 910m long. Ontario has almost no karst in its north, though limestone and gypsum are hidden beneath the muskeg. North of Lake Ontario, Moira Cave is a maze 2km long in a riverbank, and the Niagara escarpment has sinks and karren on its dolomite, but only a few short caves of mainly tectonic origin.

North of the featureless prairies there is a doline karst in boulder clay covering gypsum beside the Slave river; dolines are up to 100m across, and one river sinks, but no caves are known. The northern islands, notably Somerset and Victoria Islands, do contain extensive limestone and dolomite but the permafrost restricts caves in favour of surface ravines. Further west the South Nahanni River has cut deep canyons, leaving limestones pocked with cave entrances high in the cliffs; the longest caves are Grotte Mickey (2,270m) and Grotte Valérie (1,900m), both fossil but the latter with spectacular ice decorations. The nearby North Karst is renowned for its towers, poljes, canyons and mosquitoes but has yielded no major caves. Even further north, the Mackenzie Mountains contain huge areas of lapies on dolomite, but have no caves; there is further karst in the northern Yukon, with short frozen caves.

The Southern Rockies have the best of Canadian karst and caves, and just to their west lie the Selkirk Mountains. These have thin limestone bands with a few caves and some impressive underground drainage. The Tupper Glacier meltriver passes underground for 2km, dropping 490m, in the incredibly short time of fifty-three minutes, but the cave is unentered. Nakimu Caves are by far the longest explored (5,900m, −270m), with a complex fossil system traversed by an awe-inspiring river passage. At the north end of the Rockies, the karst of the Caribou Mountains contains numerous shafts and at least one major river sink.

Though limited in area, the limestones of Vancouver Island have yielded some fine caves, and there is still considerable potential within the wild, forested karsts. The limestone outcrops through the north end of the island contain a number of short but large river caves, some of which can be followed from sink to rising except when they are jammed by timber washed in by floods. Near Tahsis lies Thanksgiving Cave (3900m long, −223m) which contains two long streamways. Further north, the deepest caves on the island, Arch Cave (3800m long, −302m) and Glory 'Ole (1,470m long, −301m), are both single-streamway systems. In the Gold River area, a karst plateau rises 760m above the Quatsino resurgence, though the full depth potential is not yet realized. The Q5 cave reaches a depth of 287m with 2,500m of passage. Various smaller limestone outcrops around the south of the island contain some caves but may reveal many more.

Cave exploration in Canada started seriously around 1964, with mainly British cavers connected with the McMaster University group. They explored most of the main Rockies caves, but since then various provincial groups have started and continued to make new discoveries.

Southern Canadian Rockies

Bands of limestone form extensive but discontinuous karst down the axis of the great Rocky Mountain chain. High limestone escarpments, many still actively glaciated, are partly drained underground, and there are sinkholes, springs, pavements and isolated caves throughout the range. The climate and poor access restrict exploration, and there is still great potential, particularly towards the north.

Mount Robson contains the long, inclined streamway of Arctomys Cave (3,500m, −536m), and there is extensive karst northwards to the Williston Lake area, where White Hole is a shaft system 253m deep. Limestone forms impressive scarps around Jasper; underground drainage is extensive in the Maligne Canyon area, but no caves have yet been entered, while further east Cadomin Cave has a small lower passage reaching −220m. Not far south, the Columbia Icefield has some large glacier caves in its outlets, and adjacent to it is the spectacular glaciokarst of Mount Castleguard, with Canada's longest cave in the mountain of the same name.

West of Calgary, limestones form more impressive ridges and scarps, with isolated areas of karst containing dolines and sinkholes. No long caves are yet known, but the area does contain the spectacular ice caves of Plateau Mountain and Canyon Creek. The southern end of the Canadian Rockies contains more karst, including two important areas. Top of the World is a splendid level karst plateau pocked with shafts where long-lasting snow-plugs hinder exploration. Just to the east, straddling the watershed above the Crowsnest Pass, the Andy Good Plateau is a magnificent alpine karst. It has sinks, springs, lapies and dolines among the snowfields and contains various segments of fossil cave, some adorned with permanent ice. There are two major cave systems: Yorkshire Pot (6,500m long) has a shaft system descending 200m to a complex of fossil galleries reaching a depth of 384m, and Gargantua (5,800m long, −291m) has an inclined maze of passages including some very large tubes and chambers.

CASTLEGUARD CAVE

This cave is the only one in the world yet explored beneath an active ice-sheet; it is consequently of major importance with reference to sub-glacial cave and karst development. It extends beneath the Columbia Icefield, on the crest of the Rockies, and passes beneath Mount Castleguard to its entrance east of the icefield. The mainline passage is over 8km long, with the entrance at its lowest point; it is a fossil tube with a fissure incised in much of its floor, making the journey in an arduous traverse.

Castleguard was first entered in 1967, and the end of the trunk passage was reached in 1970 – at a wall of ice which is the bottom of the Columbia Icefield. Summer meltwater restricts explorations to winter; over 18km of passages have now been mapped, rising to 371m, and there is still major potential for further discoveries.

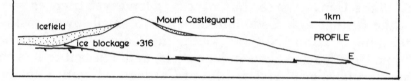

Icefield   Mount Castleguard   1km

ice blockage +316   PROFILE

E

CANARY ISLES

The seven main islands of this group are volcanic in origin and do not include any carbonate rocks. Lava tubes, however, are widely found and are frequently both deep and long. The depth in all cases is due to the surface slope rather than to any inward depth. Notable systems on Tenerife include the Cueva del Viento, whose longest unbroken segment is the Cueva de los Brevistos (7,690m long, 217m deep). On Lanzarote, the Cueva de las Verdes has 5km of tube including a chamber large enough to hold concerts in; its downflow continuation, beyond the Jameos del Agua, has been explored for 1,400m in a flooded lava tube beneath the sea floor.

CAPE VERDE ISLANDS

Of no speleological interest.

CAROLINE ISLANDS

Lying in the western Pacific Ocean, this is a widely spaced group containing over five hundred coral atolls and five larger volcanic islands, on many of which karst is recorded. The islands of Palau have the most extensive limestone, with impressive tower karst which is partly drowned; there are many large caves and Blue Holes. Ponape also has some short lava caves.

CAYMAN ISLANDS

Caribbean coral islands without speleological interest.

CENTRAL AFRICAN REPUBLIC

Of no speleological interest.

CHAD

No areas of limestone are known, but small lava caves exist on the Tarso Toucide plateau.

CHILE

The 4,000km length of this volcanically active country contains little to interest the speleologist except at its very southern tip. Here, in the centre of the inhospitable Magallanes province, are to be found two impressive but short caves in conglomerate. The Cueva Mylodon is 190m long, up to 100m wide and 25m high whilst the neighbouring Cueva Chica measures just 74m long. No records of volcanic or glacier caves are known but it seems inevitable that such features exist, particularly towards the south.

CHINA

The karst of China covers well over a million square kilometres – about one eighth of its vast land area. This has developed under a huge range of climatic controls, and limestones occur from sea-level to the heights of Tibet (and even to the summit of Chomolungma – Mount Everest). Consequently the variety of karst landscapes is unsurpassed and includes features famous around the world, notably the spectacular tower karst of Guangxi.

China's history of cave and karst exploration goes back many hundreds of years. With such a vast population living on limestone, there are now huge numbers of engineers and scientists working on the problems of karst, and local people have been into nearly all the walk-through caves. In 1981 the Institute of Karst Geology was established in Guilin. However,

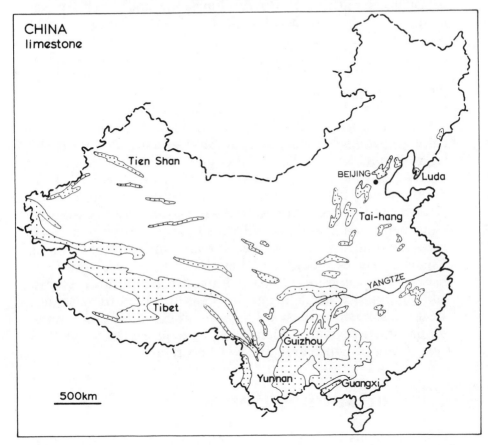

cave exploration as a sport is almost unknown, and information on cave discoveries is very patchy; the first major systematic exploration and survey were by a British expedition in 1985. Clearly the cave potential of the country is enormous, and much more is likely to be known as the political climate improves over the next few years.

Huge areas of limestone in the west have, in general, only poorly developed karst landforms. The cold, arid mountains and plateaux of Tibet have dry valleys, deep gorges and springs, but there are few known caves. There are exceptions:

the Chinghai mountains of Gansu have some areas of good karst with many dolines, shafts and caves, some of which have been explored for water resources. Caves and potholes are also recorded in the Tien Shan mountains of Xinjiang.

The limestone regions of eastern China are dominated by the great plateau of Yunnan and Guizhou, and the tower karst of Guangxi. Further north the limestone and karst are less continuous and less developed. The famous Yangtze River gorges cut through high limestone plateaux, but few caves have yet been recorded in this very extensive karst. North of the Hwang Ho, the Tai-hang Shan is a limestone plateau cut by deep gorges with many caves and springs; the Yukang caves are noted for their paintings. Many caves further west are merely rock shelters cut in the loess terraces. Toward Peking (Beijing), the karst near Shijiazhuang drains to the large Niangziguan rising, and though the limestone mountains continue further north, few caves of any size are recorded. At Zhoukoudian there are only sediment-filled fissures which yielded the famous remains of Peking Man. In Liaoning, the Luda peninsula has karst with many dry valleys and small caves, and just north of the Korean border, the caves of Xajiaweizi and Shuidong both have over 2km of passages.

Beside the limestone karst, China also has well-developed salt and gypsum karsts in several areas of the north-west, and there is a sandstone karst at Qingyan, in Hunan. Basalts in the Wudalainzhi region of Manchuria contain a number of lava caves, some of which are quite well decorated.

Yunnan Guizhou Plateau

This enormous karst plateau, hundreds of kilometres across, embraces the whole province of Guizhou and a significant part of Yunnan, and also overlaps into Guangxi and Hunan. The limestone is almost unbroken, rising from around 1,000m in eastern Guizhou to over 2,500m in the border area with Yunnan. The whole area is a spectacular karst landscape with endless ranges of cones, towers and hill peaks, and there are vast numbers of dolines, poljes and river sinks between the karst hills.

Hundreds of caves have been recorded in Guizhou, but almost all underground exploration is uncompleted. The greatest cave potential is in the marginal areas of the plateau where caves drain into deep canyons in areas of locally enhanced relief. North-west of Guiyang, Shen Dong has a single unbroken shaft of 275m, and the Daji cave has a series of large chambers beautifully decorated with towering stalagmites. Even further north the Bijie cave has a massive stream passage. In the south of the province, the Tunnanghe caves also have huge streamways, and Longgong Dong is the resurgence cave for a whole network of underground streams which drain through mainly short caves in a spectacular area of karst hills.

The higher parts of the plateau, in western Guizhou and eastern Yunnan, have the greatest depth potential for caves, and some major rivers are known to sink at considerable distances from their resurgences. The San Cha River passes beneath a large natural bridge, and then downstream drains through a deep, blind gorge into a lofty river cave over a kilometre long. The Yunnan province is also well known for its spectacular areas of pinnacle karst known as Stone Forests, near which are more sinking rivers.

Guangxi

Immediately south of the Guizhou plateau, the lower elevations of the Guangxi plains and hills are carved into the world's largest and finest area of tower karst. This covers much of Guangxi province and overlaps into Hunan, Guangdong and Jiangxi. It includes the famous area between Guilin and Yangshuo where the towers rise with vertical walls hundreds of metres high, and the Li Jiang provides one of the world's great river journeys.

Caves are known throughout the karst. Some, such as the show caves at Guilin, are restricted to individual tower hills, but even there the Qi Xing Dong has a large fossil trunk passage over a kilometre long; other caves are more extensive, and many have spectacular calcite decorations. South of Guilin a major stream flows underground through the finest of the tower karst to resurge on the banks of the Li river; the Nanxu sink cave, the intermediate Xiaoheli cave and the Guanyan resurgence system each have around 3km of passages including some very large, well-decorated streamways. Large stream caves are a feature of the area east and south of Guilin.

West of Luizhou there are many caves and underground rivers in the karst on both sides of the Hongshui river. The Soliao cave is 7,600m long and 145m deep, with a large main stream passage 4km long from entrance to sump, and east of the Hongshui a 9km-long cave has been mapped within the Tisu catchment area. Though many underground streams and rivers are utilized as water supplies, serious cave exploration has hardly started in most of the Guangxi karst.

COLOMBIA

Only three regions of karst are currently known within this little-investigated country. These all occur within the boundaries of the Cordillera Oriental, the most important and largest region being found in the province of Santander and in particular around San Gil and La Paz. Whilst both caves and potholes are to be found, it is for the latter that the area is renowned. The most spectacular of these is the Hoyo del Aire (La Paz) which consists of an entrance shaft measuring some 160m by 80m, and 187m in depth; this then leads to a very large descending passage eventually terminating at −270m. In the same area, the Hoyo del Aguila has been explored for 14km, to a depth of 217m. Many horizontal caves are situated by San Gill, with the Cueva Gdynia-Paramo (1,200m) being the longest.

A sharply eroded region of sandstone occurs by the boundaries of Cundinamarco and Tolima. Several caves have developed, but only one has been found to reach any reasonable dimensions – the Cueva de Cunday (−160m, 850m long). Slightly further south still, in Huila, a small area of karst has yielded the 823m Cueva de Guacharos.

Potential within this country cannot be considered great, but vast regions still remain almost totally unexplored, such as the densely forested east and inaccessible northern border with Venezuela.

COMORO ISLANDS

The main islands (off the coast of south-east Africa) are volcanic with no limestone. Just to the north, Aldabra atoll has karst on its coral limestone but no significant caves.

CONGO

Only one small region of karst is known within the country, situated on the Monts de Banfor. Several small caves have been explored, the longest of which is the 980m Grotte de Meya-Nzouari.

COOK ISLANDS

The islands of this Pacific chain are all volcanic. On the larger islands, rims of uplifted limestone have many short stream caves taking drainage down to sea-level sumps.

CORSICA

This large island is composed mainly of igneous rocks. Small outcrops of limestone do occur in the north-eastern highlands, where the principal systems are the Gouffre de Ghisoni (−129m and 270m long) and Gouffre de Lainosa (−80m). There is also some limestone on the Cap Corse peninsula.

COSTA RICA

The only known karst region of this predominantly volcanic country is found on the limestone of the Nicoya peninsula and in particular on the Montañas Barrahonda. Of the several systems known, the deepest is the 172m Cueva Santa Anna. The inhospitable eastern region has been little investigated, and no information is available on the wider distribution of limestone.

CRETE

Almost all the mountains composing the rugged 7,000 sq km backbone of this island are of limestone. The principal ranges are the Levka Ori, Idhi Oros and Dhikti Oros. The mountains are extremely dry but contain extensive karst features throughout. Karren, in all its forms, is abundant, while dolines, dry valleys, depressions and shafts are common. Poljes also occur. The only region which has received extensive study is that of the Levka Ori, where over two hundred caves are now known. These include both the longest and deepest systems – Mavro Skiadi (−342m, all in one shaft) and Tzani Spilios, (−280m, 2,900m long).

Local cavers are active but most major exploration has been made by French and British expeditions.

CUBA

Around seventy per cent of this large and spectacular island is formed of limestone, and much of that is well-developed cone and mogote karst with an abundance of caves. The most extensive caves are in the western province of Pinar del Río, where the Organos and Rosario mountains are steep, forested limestone; cones and mogotes rise above deep dolines and cockpits, and the ranges are separated by broad poljes. Around Vinales, the finest limestone towers of the Sierra de los

Organos contain massive caves. The Gran Caverna de Santo Tomas has 32km of passages on six connected levels, the upper ones well decorated and the lowest containing five long river passages. Nearby, the Sistema Majaguas Cantera has over 16km of passage with a long river gallery and many large side passages. In the Sierra del Rosario, the Sistema Perdidos is a resurgence cave with over 20km of mapped passages.

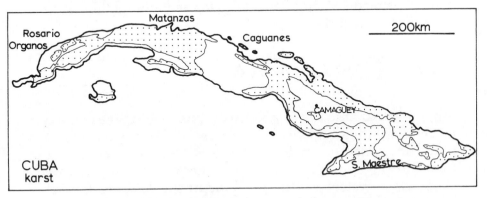

East of Havana, the Matanzas karst contains more long caves, including Cueva del Gato Jibaro (11km). But most remarkable in this area are some of the cave calcite formations: Cueva de Bellamar has fabulous calcite crystals over 50cm long in drained cave lakes, and the Santa Catalina cave contains cave mushrooms over a metre high, formed entirely of piled-up sunken calcite ice.

The Sierra Maestre, in eastern Cuba, is ringed by limestone ranges with many deep gorges and dolines, and also the island's deepest caves. Cueva Jibara descends to −248m with a number of fine waterfall shafts, and the Furnia de Pipe has a shaft 145m deep though it reaches a depth of only 165m.

Around Camaguey there are areas of deep cone karst which remain hardly explored for caves. Elsewhere caves are known in many smaller regions both on the mainland and on the smaller islands; the Cayo Caguanes is small but has over 11km of mapped caves. There is still considerable potential for new exploration. The science, exploration and conservation of caves are taken very seriously in Cuba, with a central institute and caving groups in every region, but access by foreigners is not easy.

CYPRUS

The Kyrenea and Kapas mountains, which lie alongside the narrow northern coastal plain, are formed principally of massive Cretaceous limestones. Minor caves and other karst features are known throughout, with the most interesting regions being on Mount Pentadaktylos and around the villages of Halifa and Agirda.

CZECHOSLOVAKIA

More than 3,000 sq km of karst are known within the country, in three different regions. To the west, the Bohemian karst is found in relatively small areas and is frequently covered in forest. Springs and depressions are common features. The one major cave system is that of Konepruske Jeskyne (2,050m long), south-west of Prague. Situated in the centre is the Moravian karst, again developed within several areas but with each being much more interesting and extensive. Speleological research has been pursued intensively for several decades, but even so, new discoveries and extensions continue to be made. Several long caves exist here, the greatest being the 32,500-long and 191m-deep Systema Amaterska/Punkevni, just north of Brno. Recently divers have connected the Byci Skala to the Rudice Ponor to make a system 12km long.

Almost ninety per cent of the karst and eighty per cent of the caves are to be found in the third region, the Carpathian mountains of Slovakia. The karst is varied and mostly well developed. In the alpine regions of the Mala Fatra, Belansky Tatry and Nizky Tatry, it is particularly impressive, with exposed limestone containing karren, dolines, shafts and many major caves such as Stary Hrad (−427m, 4.5km long) and the Systema v. Zaskoci/Na Prednych (−284m, 5km long). The Nizky mountains also contain the many caves of Demanova, two of which have over 8km of passages. An important division of the Carpathians is the Slovensky kras to the west of Kosice. Plateau karst predominates. Caves and potholes are common, the principal ones being the Systema Brazda/-Barazdalas (−179m) and Domica, the Czechoslovakia end of the Hungarian Baradla cave extending for an international total length of around 25km.

Caving dates back to 1748, when a descent into the impressive Macocha shaft in Moravia was made. By the start of this century serious speleological work was being undertaken, and it has continued until the present day.

DENMARK

Chalk is to be found in the northern part of the country but is of little speleological interest.

DJIBOUTI

Of no speleological interest.

DOMINICAN REPUBLIC

In addition to possessing the highest volcanic range in the West Indies (Pico Duante, 3,175m), Dominica is part of an island whose limestone content is second only to that of Cuba, occurring in five major regions. While karst is known to be

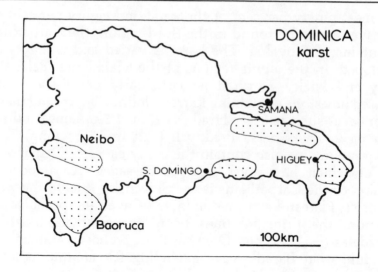

extensive, with potential for both long and deep caves, no serious work has yet been undertaken. The most promising range for caves is perhaps the Sierra de Baoruca situated between the Haitian border and Barahona. Dolines, sinks, poljes and dry valleys have all been recorded. To the north of the Lago de Enriquillo lies the Sierra de Neibo, again limestone but densely forested and little known. The third region is to the south of the Bahía de Samana, where large cave entrances, resurgences, cone karst and an arid landscape are recorded. At the eastern tip of the island, by Higuey, is a low-lying plateau with potential for cave-divers, and a similar area to the east of Santo Domingo has already received visits, with underwater caves up to 500m long (Los Manantiales) being found.

DUTCH ANTILLES

Areas of limestone are known on all the major islands, Aruba, Curaçao and Bonaire, but references to minor karst features are recorded only for the former two. The caves of the north coast of Aruba are noted for their considerable bat populations.

EASTER ISLAND (Isla de Pascua)

Isolated in the east Pacific, Easter Island has several lava caves on the slopes of Terevaka, the longest, Ana Te Pahu, just reaching 800m. The deepest is Ana Oke Ke, (+40m, 450m long). Several sea caves are also known, and depressions, springs and small caves have formed in one limited area of limestone.

ECUADOR

The considerable areas of karst to be found within this tropical country are all situated within the oriental provinces of Morona Santiago, Pastaza and Napo. In all cases exploration is

made difficult by dense forest or poor communications, or both. The greatest extent of carbonate rocks is found within Morona Santiago in the Cordilleras Condor and Cutuca. Limited exploration has revealed the Cueva de los Tayos (186m deep, 4,900m long) and the Cueva de Yaupi (1,250m long). In the western half of Napo are two extensive regions. The northernmost, almost reaching the Colombian border, remains uninvestigated, but the area centred around Tena has yielded several fine river systems such as the Cueva de Lagerto (2,000m long) and the Cavernas de Jumandi (1,900m). In the same area, the San Bernardo and Eturco caves are each over 2km long. Further potential for long horizontal caves systems is thought to be considerable in all regions.

EGYPT

Limestone is extensive throughout the northern half of the country including the Sinai. It forms the plateaux on either side of the Nile northwards from Isna and also occurs widely in the Western Desert. Minor areas of karst are known but are not considered of interest to speleologists.

EL SALVADOR

No areas of karst are known but small lava tubes have been reported from several of the active volcanoes.

EQUATORIAL GUINEA

Of no speleological interest.

ETHIOPIA

Vast areas of little-explored limestone dominate the eastern landscape of this rugged country. In the provinces of Harar, Bale and Sidamo are over 100,000 sq km of carbonate rocks out of which barely fifteen per cent have been visited by any speleologist.

In the dry Sidamo region many karst features have been noted, including large dolines, depressions, small shafts and minor caves. Slightly further north, in the vicinity of the Webbe river and Ginir, are several of the country's longest systems, including the 2,500m-long Nur Mohamed and also Sof Omar. This latter system is 15,200m long and consists of massive streamways, large chambers and series of complex

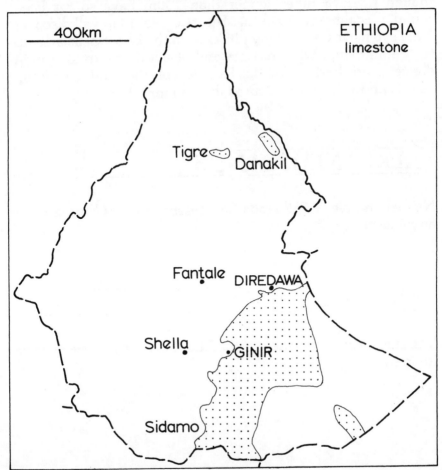

interconnecting passages. Further potential is considerable, but inaccessibility makes exploration difficult. The northernmost visited region of this block is to the south of Diredawa. Very extensive areas of karst are on record, and these contain karren, dolines, shafts and caves. The latter are to be found around the villages of Chelenko Lola, Lao Goda, Mendisa and Serkema, while shafts and the country's deepest system, Enkoftu Mohu (−192m), are close to Bedenno.

The limestone of Tigre province is around the town of Makale in an area of spectacular gorges. Karren and caves are common, although the longest system so far explored, Zayei, extends to only 300m.

Much of the Danakil mountains of Eritrea is limestone but it is not known if karst is developed.

There are volcanic areas throughout the country, and though no lava tubes longer than 100m have so far been found, it is almost certain that future exploration will discover more extensive ones. Many pits exist in the Lake Shella area up to 30m in depth. Deep fissures and blister caves up to 50m in diameter are known on the Fantale volcano, and a bedding cave is known near the Debra Libanos monastery.

FALKLAND ISLANDS

Neither the bleak Falklands nor nearby South Georgia have any limestone.

FAROE ISLANDS

Volcanic islands in the north Atlantic of no speleological interest.

FIJI

Whilst the majority of these islands are of volcanic origin, many also contain lesser areas of limestone in which karst and many hundreds of small caves have developed. The major caves are to be found on the island of Viti Levu (Wailitoa Cave, 1,400m long; Udit Cave, 800m long). Volcanic caves are so far known only on the island of Taveuni, where Salialevu Cave is 920m long.

FINLAND

Thinly banded limestone is to be found extensively throughout the country, but karst features, such as lapies, dolines and very small caves, are known only in the Lohja area to the west of Helsinki.

FRANCE

Karst covers more than 170,000 sq km (twenty-eight per cent) of the land area of France. It occurs in an exceptional diversity of forms, having developed under varying climatic conditions and at all altitudes up to 2,500m. Notable cave systems can be found in no fewer than seventy-eight of the country's ninety-three *départements* and of the remaining fifteen, most have caves of recordable dimensions. Almost twenty thousand caves are known, but most really interesting karst and all the major systems are situated within three main areas, the Alps, the Massif Central and the Pyrenees. Other important areas do exist, including, for example, the Côte d'Or, where cave-divers have gained access to many extensive systems, such as the 14,200m Grotte de Neuvon and the 13,000m Gouffre de la Combe aux Prêtres. The *département* of Meuse has the Resurgence du Rupt-du-Puits (10,250m long); Deux-Sevres contains the 4,770m Grande Fontaine de Saint-Christopher-

200km

Eure

Meuse

Yonne

Côte d'Or

Deux
Sevres

Jura

FRANCE
major
cave areas

Charente

Massif
Central

Périgord

Grande
Causses

Ardèche

Quercy

Alps

M. Noir

Pyrenees

Arbas

P.S.M.

Ariège

sur-Roc, and Yonne and Eure have a number of very different
and fascinating chalk caves.

With such vast areas of limestone, it was not surprising that
the French developed an interest in caves and karst at a very
early date. As long ago as 1765 France could claim a cave, the
Grotte de Miremont, surveyed for the incredible distance of
4,230m. With the emergence of the indomitable Edouard
Martel in the late 1800s and his exploration of more than
fifteen hundred caves throughout Europe and the United
States, the country's reputation of being the world's cradle of
speleology was assured. The pace he set then continues
unabated today. The sport and science of caving are followed
by a large number of people, and consequently there is an
almost unparalleled number of local clubs.

Alpine France

The Alps, with their infinite diversity of mountain scenery, also provide the world with one of its most spectacular cave regions. At the heart of the French zone lies Martel's 'country of the lapiaz' (lapies), an area including the sharply contrasting limestone massifs of Chartreuse, Vercors and Devoluy, all homes for major caves, such as the Gouffre Berger, the Dent de Crolles system and the Reseau des Aiguilles.

The northern outer edge of the region is the Jura, where relatively low, rolling hills predominate. Characteristic of the area are the many dry valleys, dolines, small amphitheatres and extensive river caves. Many of the caves have been explored by divers beyond long sumps and include the Reseau Souterrain du Verneau (32,100m long) and the Borne aux Cassots (15,630m).

Just north of Mont Blanc, limestones form the wild karst mountains of the Haut Giffre and the Désert de Platé, with great limestone pavements pitted with shafts and cave entrances. The caves are dominated by the Jean Bernard, currently the deepest in the world, the Gouffre Mirolda (−1,030m, 10km long) and the Gouffre du Petit Loir (−765m); all three follow down the dipping limestone after steeper entrance series cut through to the base of the main beds. This area has been the scene of the most spectacular explorations of the last decade and, as in most of alpine France, there is still great exploration potential.

Next to the south are the scarps of the Bornes and Aravais. More splendid lapies cover a huge area of the Parmelan plateau and again hide many shafts. Some of these lead down into the fine river passage of the Reseau de la Diau, creating superb through trips in a system 701m deep and nearly 15km long. The adjacent Sous Dine plateau has the Gouffre JP2 (−407m), and on the Leschaux plateau the Tanne à la R'Noille has a 176m shaft into a streamway which reaches −268m. The Reseau de la Tournette (7,400m) feeds the Fontaine du Paradis. Across Lake Annecy, the Bauges massif is dominated by the Margeriaz, with its many fine potholes, locally known as *tannes*. The Cochons-Froide system, now over 15km long, has long streamways which descend to −825m. Nearer Chambéry, Mont Revard is another important karst, containing the 20km-long Trou du Garde.

The centre of the karst belt is dominated by the magnificent massifs of the Chartreuse and Vercors; then the Devoluy is another wild alpine karst with many shafts locally known as *chourums*. The one really deep cave is the Aiguilles system. Away from the alpine terraine, the Plateau de Vaucluse is a rolling upland karst with many dry valleys. It contains the steeply descending shaft caves of Caladaire (−667m) and Jean Nouveau (−578m) and also the more complex Aven Autran

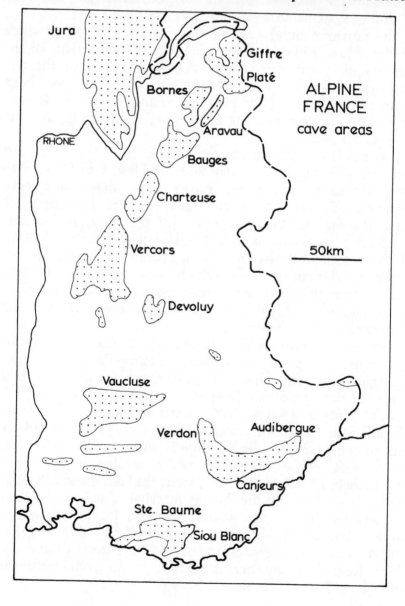

RESEAU JEAN BERNARD

Currently the world's deepest known cave system, the Jean Bernard lies with its highest entrance at an altitude of 2,264m, on the Lapies de Follis high above the Alpine town of Samoëns. There are seven entrances which lead to finely sculptured passageways steadily descending to a depth of 1,535m. The original entrance, V4, was discovered by the Groupe Vulcain in 1964. The following year a depth of −450m was reached, and by 1975 an upper entrance, B19, was found and a syphon at −1,298m achieved. Intense exploration over the following few years yielded even higher entrances, and three syphons were passed to give the current depth. Apart from a 136m-deep shaft in the B21 entrance series, there are no pitches of any size. The main hazards of exploration are the cold and the water, and the tediousness of the small canyon passage down the lower part of the cave. The total length of the system now exceeds 17km.

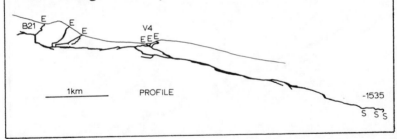

Broken karst massifs follow close to the Mediterranean coast, forming low plateaux with doline fields, dry valleys and numerous caves. Near Marseilles, the submarine resurgence of Port Miou has had its flooded cave explored for 2,200m, and not far inland the Sainte-Baume and Siou Blanc karsts both have caves over 300m deep. The Verdon gorge is cut deep into limestone but has no major caves within it, while the adjacent Canjeurs plateau has its Gros Aven, explored to −290m. On the Audibergue ridge, the Aven des Tenèbres is a shaft system reaching −440m, and the small section of the Marguareis lying in France (most is in Italy) contains the Gouffre Pentathol, whose sequence of shafts is 500m deep.

RESEAU DES AIGUILLES

The Massif du Devoluy, situated on the edge of the French Alps, contains many interesting systems but none to rival the 6,100m length and 980m depth of the Reseau des Aiguilles. This lies beneath the northern slopes of Mont Aiguille and has two entrances, the Chourum du Rama (altitude 2,273m) and the Chourum des Aiguilles (altitude 1,995m). The two systems connected at the 500m level and continue as one passage to the terminal syphon. Exploration is quite straightforward, but there are several unpleasant boulder chokes with some loose rock.

The Aiguilles entrance was initially investigated in 1965, and the bottom, at −682m, was reached in 1969. In 1972, after much arduous upstream exploration, the Rama entrance was opened from within. The water resurges at the Source des Gillards (altitude 860m).

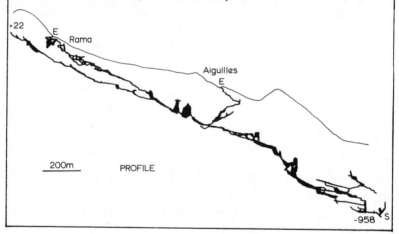

The Chartreuse Massif

The dramatic cliffs which rise to a height of 2,000m, immediately west of the Isère river above Grenoble, guard a karst exceptionally rich in caves. More than six hundred caves have so far been recorded within the two mountain chains which form the massif. The eastern chain is basically one long syncline with Mont Granier at the northern end and the Dent de Crolles to the south. Its barren surface, occasionally

RESEAU DE LA DENT DE CROLLES

This popular system is situated within the spectacular karst massif of the Dent de Crolles, in the Chartreuse, just west of the main French Alps. It comprises several inclined levels of often complex passageways connected by many shafts. The cave may be entered at any of five places, the plateau shafts of the Gouffre Thérèse and the P40, the marginal cliff entrances of the Trou de Glaz and the Grotte Annette, and through the large open resurgence of the Guiers Mort. Its popularity stems from its ease of access, its considerable through-trip options, its fine passageways and shafts, its complexity and its generally challenging nature.

The major explorations took place between 1936 and 1947 by the team led by Chevalier and Petzl. These involved the most spectacular caving of their time, notable for the fact that much of the exploration was carried out upwards from the Glaz to the P40. Since then there have been regular extensions; notable were the connection from the Thérèse entrance, taking the depth to 623m, and the discovery of the great tunnels of the Métro level. The total length is now 53.8km.

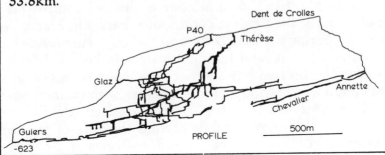

enlivened by a stunted tree, is well known for its extensive fields of lapies, dolines and shafts, while the cliff faces and lower slopes contain resurgences and open caves. Two major systems dominate the range, the Reseau de la Dent de Crolles and, further north, the Reseau de l'Alpe, a complex cave with twelve entrances and a warren of active and fossil passages extending for more than 51km with a vertical range of 602m.

The western chain is broken by the gorges of two rivers, the Guiers Mort and the Guiers Vif, and is geologically much more complex and varied. Mont Outheran and the Grand Som are both rugged and wild, the former containing many minor systems and the latter the major Puits Francis (−723m) and the Puits 508 (−481m). The Charmant Som is formed on an anticline, and to date the only known system is the steeply descending, difficult Reseau Ded (−780m). The southernmost and rather lower block is that of Genieux; it is extensively wooded but many small caves have been found, together with the shaft system of the Scialet de Genieux (−675m).

Martel, as usual, was the first serious explorer of the massif, making a visit to the Trou de Glaz in 1899. Both de Joly and Chevalier considerably extended his work, and by 1960 systematic exploration had begun. Such intensive studies over the past twenty years make further major finds unlikely, but the western mountains could still hold some surprises.

Vercors

This spectacular block of limestone is situated south-west of Grenoble. It is the largest continuous karst in France and comprises various plateaux almost wholly surrounded by precipitous cliffs several hundred metres in height. Much of the surface is forested but areas of bare lapies do occur, as on Sornin and Grande Moucherolle, together with dolines, shafts and deep potholes. In the clearings, deep dolines and depressions are common, while around the base of the cliffs are many very fine river caves and resurgences. Several of these have been explored for considerable distances upwards: the Grotte de Gournier (+680m, 15,120m long) has a well-decorated fossil passage with a splendid streamway beyond, while the adjacent Reseau Coufin-Chevaline (+401m, 15,060m long) rises behind its spectacular show-cave passage. Deep caves occur on the high plateaux, which may be up to 1,400m above the valley floors. Notable systems include the classic Gouffre Berger, the difficult Scialet de la Fromagère (−902m) and the 580m-deep complex of the Scialet Combe de Fer. Also worthy of mention are Le Pot II, a single shaft 337m deep, and the recently explored Antre des Damnés above Correncon, which is now 723m deep.

GOUFFRE BERGER

Perhaps the most famous of all the world's caves, the Berger has been known since 1953 and became the first system explored beyond the elusive 1,000m depth in 1956. Its depth remained unsurpassed for twelve years. The Berger is situated amongst the superbly wooded lapies of the Sornin plateau at the northern end of the Vercors massif. Four entrances are now known, the Gouffre Berger (altitude 1,460m), the Puits Mary (altitude 1,440m), the Gouffre des Elfes (altitude 1,425m) and the most recently discovered Gouffre des Rhododendrons (altitude 1,510m). These inlets each descend in a rapid series of shafts to join the main drain at around the 300m level. Enormous galleries lead through even larger chambers, with a steady descent to the final wet pitches to the syphon. Six sumps have been dived to give the present depth of −1,248m and length of 24,400m. The water resurges at the Cuves de Sassenage (altitude 297m) just over 4km away. A fifth system, the difficult Scialet de Fromagère (−902m) drains into the lower end of the Berger, but so far no physical connection has been made.

In spite of having been known for so long, the system is still commonly regarded as the finest caving trip in the world. It embodies everything that is good and exciting about caving, with its clean shafts, large passageways and continuous descent. The impressive size of the Grand Eboulis (Big Boulder Pile) and the superb formations of the Salle des Treize (Hall of Thirteen) are followed by the canyons, canals and cascades down to Claudines, and the final series of wet pitches, climaxing in the Puits de l'Ouragan (Hurricane Shaft).

Cave explorations in the Vercors were first made around 1900 by Martel and Decombaz. The Grotte de Bournillon, with its enormous entrance porch, and the Grotte de Brudour (now 8,310m long) were early finds. Many minor discoveries were made over the next fifty years, but was not until systematic exploration started in 1952 that any deep caves were found. Early discoveries included the flood-prone Grotte de la Luire (−451m, +63m), the Sassenage resurgence cave, the Gournier and the Berger. There still remains considerable potential for new systems and extensions to old ones.

Massif Central

This large region of central and southern France is composed mainly of granites, gneisses and volcanics. Limestones occur chiefly around the edges and have developed into several areas of major importance. These are Charente, Périgord and Quercy in the west, Montagne Noir and the Grande Causses in the south, and along the eastern margins within Ardèche and Gard. Over six thousand caves have so far been recorded, and though none is of great depth, many are long and of considerable beauty.

The areas of the west comprise mainly dry plateaux dotted by circular depressions, lined by dry valleys and rich in caves and *igues* – a local term for shafts. Many of the caves are of great pre-historic importance, mainly in Périgord (Grottes de Lascaux, Font-de-Gaum, Pêche Merle and Cougnac). In the Périgord of Dordogne, the Villars and Miremont caves are each around 7,000m long, and the resurgence of Doux de Coly has recently been dived for 3,100m. The major cave systems are found within Quercy, and in particular on the Causse de Gramat. Most notable here is the spectacular shaft and river passage of the Gouffre de Padirac (230m deep, over 17km long). Charente, while containing extensive areas of limestone, can lay claim to only one long cave, the 9,380m Reseau du Trou qui Fume.

The Montagne Noir, with its many small cave systems, and the adjoining Grande Causses form the southern boundary of the Massif. The Causses are scenically impressive, being a series of arid tabular blocks (Sauveterre, Sévérac, Méjean, Noir, Larzac etc), separated by deep gorges (Tarn, Jonte,

Dourbie, Vis etc). Small trees and shrubs give a wild beauty to karsts of lapies and multi-shaped dolines; below ground lie some of the country's most beautiful caves, culminating in Aven Armand (Mejean) with its towering stalagmites.

The eastern edge of the Massif stops abruptly above the Rhône river, and in the karst areas of Ardèche and Gard almost three thousand caves have been explored. Undulating and often wooded plateaux are again separated by deep gorges. Caves, lapies and dolines are everywhere, with the caves usually wetter and longer than those in the west; they include the Grotte de Saint-Marcel d'Ardèche (24,750m) and the Reseau de Foussoubie (22,750m long).

French Pyrenees

The Pyrenees form a geologically complex mountain chain 400km long between the Bay of Biscay and the Mediterranean. Limestone karst is found in a broken belt extending through almost the full northern length of the mountains, and in the higher massifs of the west and centre. It is in these latter regions that the greatest caving areas are situated. Several of the world's deepest known systems exist here in a terrain consisting of some of the wildest and most exciting karst in the world. This culminates in the incredible Massif de la Pierre Saint-Martin, which also extends, as Larra, into Spain. Pride of place here goes to the Gouffre de la Pierre Saint-Martin (1,342m deep). The nearby Massif d'Iseye, not quite as rugged but equally cavernous, houses four major systems, including the Gouffre du Cambou de Liard (926m deep) and the Gouffre Touya de Liet. The remaining western ranges of Urkulu, Arbailles, Issaux and Ger have many caves but to date none deeper than 600m. Exploration in all these areas may be considered to be in its infancy, and other major discoveries are expected.

The *département* of the Hautes Pyrenees has no great caving regions, but special mentions must be made of the spectacular amphitheatre of the Cirque de Gavarnie, with its many small caves, adjacent to the Massif du Marbore, a major karst area of northern Spain. In Haute Garronne there is only one massif of importance, that of the Arbas-Paloumère, second only to the

RESEAU DE LA PIERRE SAINT-MARTIN

The five natural entrances of this great system are situated on the Massif de la Pierre Saint-Martin astride the Spanish/French border in the Pyrenees Atlantiques. This constantly expanding cave currently measures 50km long and 1,342m in depth. The highest and most recently explored entrance is the M31 (altitude 2,058m), and the original and most well known is the Sima Lepineux (altitude 1,717m). Entry down this 320m shaft to the enormous passageways below was first made in 1951. In 1960 a tunnel was dug by the EDF in the hope of making use of the cave river, and this gave easy access to the lower reaches via the voluminous Salle de la Verna. The SC3 was first descended in 1975. This consists of a rapid series of pitches to a depth of 360m. It is then 11,500m of variable passageway to the bottom; while there are no great technical difficulties, the very length makes such a trip arduous and rarely completed, particularly as only a miserable series of meanders and shafts continue beyond the Verna. In contrast, the passages upstream of the big chamber are mostly large and spectacular.

The cave water rises in the Resurgence de Bentia (altitude 445m) not far from the village of Sainte-Engrace. Further extensions towards this rising are thought unlikely but new and higher entrances are a distinct possibility.

The Massif de la Pierre Saint-Martin is a large and exceptional karst region. Vast barren fields of lapies extend as far as the eye can see, while jagged dolines, depressions, dry valleys, shafts and caves occur in profusion. Twelve systems, exceeding 2km in length and 385m in depth, are already known; they include the Gouffre Lonne Peyret, with its two entrances systems of shafts leading to the long river passage, finally choked at a depth of 774m very close to the Arphidia cave. There are numerous unexplored possibilities for even larger systems. The region was recognized for its unique characteristics at an early stage, and in 1965 ARSIP was formed (Association Recherches Spéléologues Internationales à la Pierre Saint-Martin). Its job was to co-ordinate all activities in the area, which it has done with success.

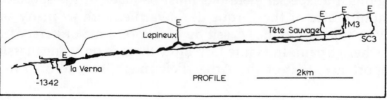

PROFILE

RESEAU DE LA COUME D'HYOUERNEDO

The beautifully wooded Massif d'Arbas in the French Pyreneean *département* of the Haute Garonne is home for one of the world's most remarkable systems. Previously known as the Reseau Félix Trombe, it currently exceeds 75km in length and is 1,004m deep. It comprises an amazing complex of both active and fossil passageways with no fewer than twenty-eight entrance systems. Some of these, such as the famous Gouffre de la Henne Morte and the Grotte de Péne Blanque, can be considered major caves in their own right. The highest entrance known so far, the Grotte de la Coquille, lies at an altitude of 1,457m, and the lowest, the resurgence of Grotte du Goueil di Her, is at 486m. Unfortunately this latter entrance can be gained only by divers capable of reaching a depth of −38m in the syphon. The potential for further extensions and connections is considerable, and French cavers are at this moment attempting to join the Système Supérieur de la Coume Ouarnède to the main system. This would have the effect of adding on a further 10km of mapped cave but would add little to the depth.

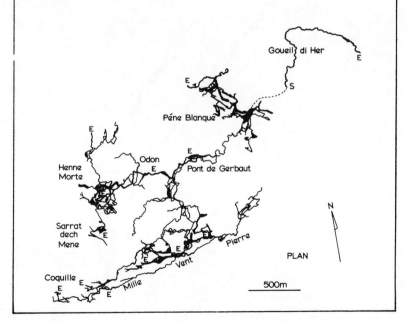

GOUFFRE TOUYA DE LIET

The extensive lapies of the Massif d'Iseye, high above the town of Accous in the Pyrenees Atlantiques, currently contains four systems deeper than 600m. The deepest of these is the staircase of pitches forming the Gouffre Cambour de Liard (−926m), but the most interesting is undoubtedly the 917m-deep and 4,250m-long Gouffre Touya de Liet, which commences with three adjacent entrances up to an altitude of 2,060m. The passage then follows the limestone bedding down an amazing series of ramps at an average angle of 40 degrees, with several minor pitches, to a depth of 610m. At this point a scramble over boulders reveals a massive but broken 302m shaft. Just prior to this pitch, a second system enters the cave, known as the Gouffre de la Porte Etroite. The cave was first discovered in 1973 and the bottom reached the following year. The water resurges in the Valée d'Aspe.

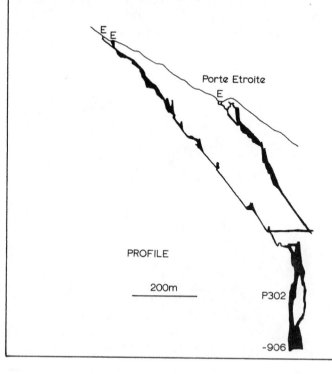

Porte Etroite

E

PROFILE

200m

P302

−906

Massif de la Pierre Saint-Martin. Within this hollow mountain there lies the deep system of the Coume d'Hyouernedo (including the Henne Morte). Many other caves exist but, not surprisingly, not of such impressive dimensions.

Ariège offers perhaps the greatest diversity of karst phenomena in a variety of settings. Lapies, dolines, poljes, sinkholes and depressions are all common, as are the caves. These include the Gouffre Georges (−694m), the beautifully decorated and well-known Grotte de la Cigalère (Haut-Lez; 9,630m long) and the Reseau de Niaux-Lombrives (Tarascon; 8,000m). Others are of prehistoric importance, and the Grotte de Moulis is an underground speleological laboratory, while the remarkable Mas d'Azil is traversed for 420m by a two-lane highway. No major cave regions exist in the limited karst of the easternmost Pyrenees although around Villefranche de Conflent several reasonably extensive systems are found. To the north, in Aude, similar karsts are known near Sault and in the Feuilla region of the eastern Corbières.

GABON

In the hills between Lastoursville and Kissipougou is an extensive region of karst recently explored by a French expedition. Several well-decorated caves are now known, including the long Grotte de Kissipougou and the 1,400m-long Grotte de Lastoursville.

GALAPAGOS

These south Pacific Darwinian islands of volcanic origin are rich in caves. They occur principally on the islands of Santa Cruz (Gallardo, 2,150m long) and San Salvador (Bucanero 1, −57m). Lesser caves are to be found on each of the other seven main islands.

GAMBIA

Of no speleological interest.

GERMAN DEMOCRATIC REPUBLIC (East Germany)

Extensive areas of both gypsum and limestone karst occur around the Thuringian Basin, situated in the south-west corner of the country. The important Harz mountains form its northern boundary. On their southern slopes, among numerous dolines, small poljes and springs, are several quite long caves situated in gypsum (Heinkehlesystem, 4km long; Wimmelburgerschlotten, 2,400m long). On the opposite side of the hills, particularly around Rübeland, are similar features in limestone, including Hermanshöhle (1,750m long).

To the south of the basin, along the Thuringian Forest and in the Vogtland, outcrops of reef limestone and gypsum support many karst features and contain many small caves.

GERMAN FEDERAL REPUBLIC (West Germany)

Considerable areas of limestone occur throughout the country, but those producing significant karst are few. Intensive investigation of these regions has yielded some three thousand caves. The majority are small, although many compensate for lack of size by being of particular beauty or containing some surprising feature. Four major caving regions can be identified, with the Berchtesgaden area in the alpine borderland proving the most important. Here are to be found all the deepest caves (ZE-Schlinger, −584m; Kargrabenhöhle, −447m; Geburtstag-schacht, −698m; Warnischacht, −550m). Wild plateaux of barren karst, such as the Gottersacher, contain extensive

karren, dolines, dry valleys, shafts and caves. The main
potential for future exploration also lies on these plateaux.

Much further north, the rolling Sauerland contains an
abundance of richly decorated and often extensive systems
(Kluterthöhle, 5,200m long), while across to the east the
Frankischeralb is Germany's largest karst area, with its
comparable continuation in the Schwäbischeralb. This covers
some 6,400 sq km and contains over thirteen hundred caves.
Unfortunately few reach any major dimensions. The final
region, and perhaps the most interesting speleologically, is that
of the Harz mountains. These are an alternating mixture of
limestone and gypsum which has produced some unusual
karst scenery. Sinks, dolines, dry valleys and spring abound,
and there are many caves. Some of these are noted for their
large chambers; in the Himmelreichhöhle gypsum cave one
room measures 170m by 80m by 15m in height.

GHANA

Of no known speleological interest.

GIBRALTAR

The Rock consists of a massive block of Jurassic limestone rising to 426m. Many small caves are known, and the largest is Old St Michael's Cave with a series of decorated chambers reaching to a depth of 76m; it is now developed as a show cave. Many caves have yielded bone material, notably Forbes Quarry Cave in which a Neanderthal skull was found.

GREAT BRITAIN

The landscapes of Britain contain some spectacular and varied karsts, which lack only deep caves, due to the limitations of relief. The most important karst is on the Carboniferous limestones – with fine glaciated pavements and scars in the north contrasting with the dolines and dry valleys of the south. Most of the caves are in four regions: the Yorkshire Dales, the Peak District, the Mendip Hills and South Wales. Elsewhere the Carboniferous limestone supports lesser karst features. Round the Lake District are spectacular limestone pavements but few significant caves, and thinner limestone bands in the northern Pennines contain isolated stream caves and phreatic mazes. In North Wales, the bold limestone escarpments contain scattered caves, including Ogof Hesp Alyn, a 2km phreatic system drained by mining activity, and the recently discovered Ogof Llyn Parc with its 3km of streamways and fossil tunnels.

Older limestone contains a number of caves in the Inchnadamph area of Scotland, where the longest cave is Uamh an Claonite, with 2,000m of passage including some large streamways. Scattered caves also occur in Devon, where the Baker's Pit system is a small-scale warren containing 3km

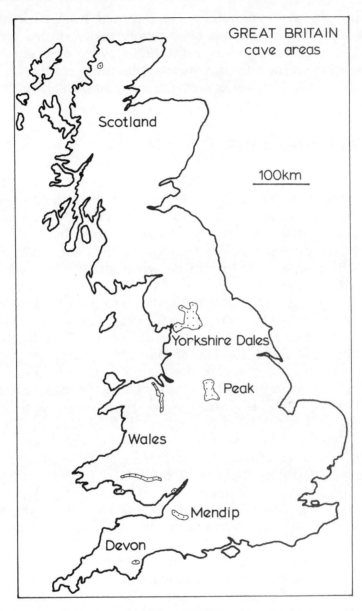

GREAT BRITAIN
cave areas

Scotland

100km

Yorkshire Dales

Peak

Wales

Mendip

Devon

of passages. The younger limestones of England have formed extensive karst areas dominated by dry valleys but there are few caves. In the North Yorkshire Moors, the Windypits are tectonic fissures up to 43m deep; the Cotswolds are essentially non-cavernous, and the Isle of Portland contains only small caves. Similarly the chalk has only isolated caves (Beachy Head Cave, 400m long, in Sussex), though its drainage is mostly underground.

Though cave explorations are recorded from back in the sixteenth century, serious caving only really started around 1900. The golden age was in the 1930s, when open entrances lay waiting on the hills, but major explorations continue today, and the rate of discoveries seems to show no slackening.

Yorkshire Dales

The thick slab of Great Scar Limestone reaches right across the Yorkshire Pennines and forms the finest karst landscapes in Britain. The limestone is cut by the glaciated valleys of the Dales, and many of the benches are covered by spectacular limestone pavements. There are hundreds of caves lying beneath sinkholes fed by streams which drain from shale hills above the limestone.

The longest caves are in the western part of the karst, around the hill of Gragareth. On its west is the enormous Ease Gill Cave System, and on its east is the West Kingsdale System; this has 11,600m of passage with three pothole entrances to a main streamway with a 1,800m-long sump to the Keld Head resurgence. In the centre of the main cave area, Ingleborough hill is characterized by more vertical caves. Gaping Gill is the most famous, with its 110m shaft into a large chamber; the system is over 14km in length and 204m deep, with a great complex of fossil passage leading to the sumps dived to the Ingleborough Cave resurgence. The great shaft systems of Juniper Gulf (128m deep), Alum Pot (−107m) and Meregill Hole (−173m) are among the most notable on Ingleborough, and White Scar Cave is a 6km-long resurgence system.

Further east, Penyghent Pot (−184m) is a magnificent system of streamways and cascades, just the finest of the many caves in the hill of the same name. The Malham area has the best of the surface karst with pavements, sinks, gorges and the famous Cove, but few caves have been explored. East of Malham, Wharfedale has some notable caves, including Birks Fell Cave (−142m, 3,600m long), the long, straight through route of Dow Cave (3,600m long) and the superbly decorated Strans Gill Pot (−103m). High on Great Whernside are the long streamways of Langcliffe Pot (9,600m long, −116m) and

EASE GILL CAVE SYSTEM

At the western end of the Yorkshire Dales (but actually in Cumbria and Lancashire) a belt of limestone fells drains underground to a single resurgence at Leck Beck Head. Nearly fifty years of continuous exploration in the many potholes has revealed a rambling but integrated cave systems containing over 52km of passages, the longest yet known in Britain.

On the north side of the Ease Gill valley numerous streamway caves drain into the Ease Gill master cave. This lies beneath a series of large decorated fossil passages, and both high- and low-level caves can be followed to beneath the Lancaster Hole entrance. Low-level passages also pass beneath the surface streambed and connect to Pippikin Hole. This has a number of small streamways cutting through a complex of large fossil tunnels. Further south, the fine streamway cave of Lost Johns is connected to the Gavel Pot system through the downstream sumps, and so has a vertical range of 186m and a length of 11km. Divers have explored the sumps of Pippikin and Gavel, and the connection is nearly made; this will create a system with a total length of over 63km. Another dive will connect the Gavel sump to the recently extended Notts Ireby system, to take the total length to over 70km.

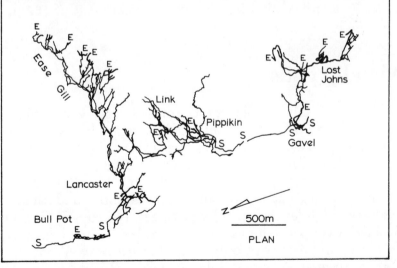

Mossdale Caverns (10km long), and below its eastern flank the River Nidd goes underground through the largely flooded tunnels of the Goyden Pot System (6,200m).

Peak District

Derbyshire has the largest unbroken area in Britain of well-developed karst, in the southern half of the Peak District – often known as the White Peak. The rolling limestone plateau is mainly covered in grassland; it is broken by numerous dry valleys, some of which are incised into more spectacular gorges near the plateau margins. Some of the gorges, notably the Manifold and Stoney Middleton, have associated cave systems of no great extent; sinks are not common, and few resurgences can be extended.

The only important area of caves is around Castleton. One of a line of sinkholes, Giant's Hole, has a long, meandering streamway in its 4,700m of passage; from the Oxlow entrance higher on the hill it is 214m deep. There is no connection passable to the resurgence complex of Peak Cavern and Speedwell Cavern. These caves, linked by fossil and flood routes, have two long, parallel streamways, totalling over 10km in length, with inlets giving a vertical range of 182m.

Mendip Hills

This small, isolated plateau is formed largely of limestone which supports a landscape of dolines, dry valleys and marginal gorges. Most caves are in the Priddy region of western Mendip. The finest cave is Swildon's, with over 8,000m of passage reaching a depth of 167m; its long streamway descends through twelve sumps, and there are extensive high levels and inlet passages. Nearby, St Cuthberts Swallet (over 7km long) has a complex inclined maze well decorated in parts with only a short streamway. Waters from both resurge at Wookey Hole where the River Axe has been explored through twenty-five sumps and underwater chambers; including dry high levels, the cave is now 3,300m long.

Mendip's deepest cave (Longwood Swallet, with rift complexes and a stream canyon reaching to −175m) lies north-west of Priddy. Its water combines with that from the well-decorated G.B. Cavern (2km long, −134m) and also Manor Farm Swallet (−151m) emerge at Cheddar, where only short fossil caves are entered in the famous gorge. A small group of caves in eastern Mendip includes the adjacent Shatter Cave and Withyhill Cave, notable for their spectacular decorations.

South Wales

Along the northern margin of the South Wales coalfield, thin outcrops of gently dipping limestone are crossed by a number of large streams. The karst landscapes are not spectacular, but they are underlain by some long and very fine cave systems. The Swansea Valley contains two superb caves. On the east, Ogof Ffynnon Ddu is the deepest in Britain, at 308m, and has over 43km of passages; it is a complex sloping maze draining to a canyon streamway 5km long. Across the valley, Dan yr Ogof is another complex fossil maze, with 15km of passage rising 140m above the resurgence entrance; though its streamways are shorter, it has some spectacular decorations.

Further east, the Neath Valley contains the Little Neath River Cave (8,200m long with some fine, large river passages), and shorter, largely flooded caves carry the Mellte and Hepste rivers. Only short caves are known between there and the north-eastern end of the limestone with its spectacular caves beneath the Llangattwg scarp. Isolated from the main cave areas, Otter Hole (3,200m) lies almost at sea-level and is entered through a unique tidal sump; beyond it a long, high-level passage contains some of the largest and finest stalactites and stalagmites in Britain.

LLANGATTWG CAVE SYSTEMS

The limestone beneath Mynydd Llangattwg all drains to the Clydach Gorge, through a series of caves which, though not yet connected, has the potential to be one of Britain's greatest systems. Ogof Agen Allwedd was largely explored in 1957, though many minor discoveries have been made since then. Some of the fossil passages and much of the main streamway are of large proportions, and the cave is now 29km long, with a vertical range of 180m.

The second phase of discoveries started in 1976, when entry was gained to the lower end of the fossil passages of Ogof Craig-a-Ffynnon. This cave now has over 8km of galleries, mostly large, dry tunnels which are spectacularly decorated in some places. Then in 1984 and 1985 Ogof Daren Cilau was extended beyond its small entrance crawl. Over 12km of passages were quickly discovered, including the main-line tunnel, which at 20m high and wide is the largest passage in Britain. There are great possibilities of connections to both the other caves.

GREECE

Almost eighty per cent of the country's mountainous regions is composed of limestones, and of these the majority exhibit karst phenomena to some degree. Serious speleological investigations, with certain exceptions, have been minimal, and the potential for further discoveries is thought to be considerable.

In the north, extensive and recently explored karst occurs in the Khalkidhiki mountains (Kokkines Petres, 1,500m long, with finds of Neanderthal man), north of Makir (Kyklops Polifimos, 1,800m long), between the Bulgarian border and the towns of Serrai and Drama (Alistrate, 2,500m of very large

GREECE
karst

and beautiful streamway) and finally to the west of Florina around Prespansko lake.

On the main peninsula the western coastal side is mainly a mixture of limestone and sandstone, whilst in the east the majority of caves are found towards the south (Parnassos and around Athens). Between the east and western regions lies a central core of predominantly limestone mountains, extending from the Albanian border to the Gulf of Corinth. This is really a continuation of the Dinaric mountains of Yugoslavia and contains the finest karst areas. The most renowned of these is within the Timfi Oros of the northern Pindhos where great blocks of limestone, such as Astraka, have been cut by deep gorges like the Vicos. The plateaux are wild and uninhabited places covered by great fields of lapies, dolines, depressions, fissures and many spectacular shafts, such as the Epos Chasm (−451m) and Provatina (−405m). Even these have barely pricked the 1,400m potential to the Vicos risings. Karst continues southwards, but the only other deep systems are found to the west of Karpenision (Agathi, −190m), while Perama is a 1,900m-long horizontal decorated show cave near Ioannina.

In the Peloponnese, the main karsts extend from the mountains of Corinth (Limnes, 2,000m long), past the west of Argos (Haghios Elios, −150m) and on to the wild Parnon Oros. Here are potholes to rival those of the Pindhos, with Propantes (−315m) and Skorpion (−209m). The extensive area around Areopolis contains the country's longest cave, Vlyhada, originally entered by boat from the sea and now 3,400m long.

Many of the Greek islands are dominantly limestone. Kefallinia has its famous seawater sinkholes, Euboea contains the 1,500m-long cave of Agea Trias, and there are recorded caves on many others.

GREENLAND

Small areas of limestone exist in the north-east, and numerous small caves have been noted. Among the innumerable glaciers of the east and west there is considerable potential for the exploration of glacier caves and moulins.

GUATEMALA

The karst of this relatively small state is of major significance due to its profusion and diversity. Over one third of the country is composed of limestones, which occur in four major regions – Alta Verapaz, Huehuetenango, Quiche and Peten. The karst varies from tropical lowland through to high alpine and in almost all cases is very well developed.

The principal region, due mainly to its accessibility, is Alta-Verapaz, extending from the Río Chixoy eastwards as far as the Gulf of Honduras. It is then bounded in the south by the Polochic river and in the north by Peten. The relief varies considerably, ranging from sea-level to over 2,500m, and the karst, which invariably occurs in dense forest, is distinctly tropical. Cones and towers abound; then there are great sinks and resurgences, dolines and some enormous poljes (Sesejal measures 45km by up to 5km wide). Caves, of which well over four hundred are known, are typified by river systems. The most notable of these is the 21,980m-long Sistema del Río Candelaria (Chisec). This is basically seven successive river caves, separated by no more than 800m, along the course of the Candelaria river. The longest single section is the Cueva Veronica del Río Candelaria (7,900m). Further to the south lies the region's deepest cave, the Sistema de Seamay-Sejul (Senahue; −170m, 3,000m long).

Huehuetenango, to the west, has a totally different setting in an area reaching up to 3,800m in height. The forests here are mainly coniferous and are found scattered between great prairies and rugged mountains. The karst is alpine in character, with great fields of lapies, dolines, shafts and dry valleys. The potential is thought considerable, and most caves so far discovered tend to be more vertical than horizontal – El Ojo Grande de Mal Pais (−240m) and El Ojo Chiquito de Mal Pais (−210m), both near Barillas. But near El Tabacal, the Cueva de Agua Escondida is a resurgence system with over 3km of large river passages, mostly negotiated by boat. The Altos Cuchumatanes bounding the south are relatively unexplored but are known to contain many caves, particularly in the Llanes area.

The border country to the west of the Río Chixoy known as Quiche is again rich in karst but, due to a dense covering of vegetation and its inaccessibility, no research has so far been undertaken.

The final limestone region is that to be found within Peten. It is of low relief, rarely reaching 300m except at the end of the Montañas Mayas, and is little visited due to its swamps, jungle and biting insects. Small caves are known, the longest being the Cueva de Jobitzina, around 1km.

GUIANA, FRENCH

There are no known areas of karst, but a small cave in laterite has been found not far from Cayenne.

GUINEA

Areas of limestone karst are unknown but small sandstone caves do exist in the Fouta Djallon massif.

GREAT BRITAIN: Gaping Gill, Yorkshire Dales

GREAT BRITAIN: Ingleborough,
Yorkshire Dales

GREAT BRITAIN:
Ogof Ffynnon Ddu, Wales

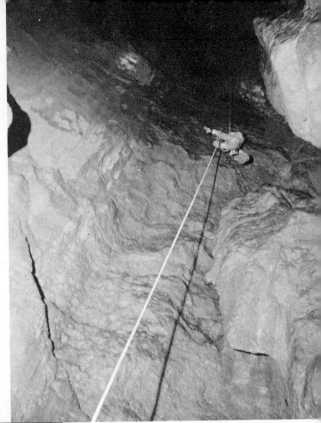

GREECE:
Tripa tis Nifis

ITALY: Antro del Corchia

HAWAII: Kazumura Cave

INDIA: Amarnath Cave, Kashmir

INDONESIA: Gua Bribin, Java

IRELAND: Marble Arch Cave, Northern Ireland

JAMAICA:
Golding River Cave

JAMAICA:
Quashies River Cave

JAPAN: Akiyoshi-do

MALAYA: Batu Temple Cave

MEXICO: Gruta del Palmito

GUINEA BISSAU

Of no speleological interest.

GUYANA

No areas of carbonate karst are known but, among the headwater regions of the Kamarang river, cave and shaft entrances have been noted in the Roraima sandstone.

HAITI

Forming the western half of Hispaniola, Haiti is extremely rich in limestones, with unexplored karst occurring in many areas. The Selle massif, between Jacmel, Port-au-Prince and the Dominican border, is an almost continuous region of lapies, dolines, poljes, shafts, dry valleys and caves. Reconnaissance visits have revealed, among others, the Trouin Sue (−91m, 1,670m long) and the Bim Sejourne (−167m). Further west the Massif de la Hotte is similarly endowed, as are the more central hills of Matheux, Noire and Nord and the island of Gonave.

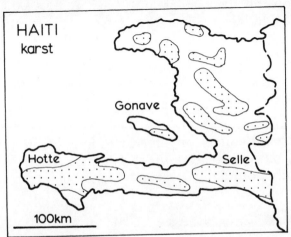

HAWAII

Created entirely by a string of volcanoes, the Hawaii islands have no limestone, but within the long basalt flows there lies a spectacular series of lava caves. The finest are on the youngest island, Hawaii itself, and the Kilauea volcano has new lava tubes forming nearly every year. Some of the youngest flows have caves notable for their lava drip formations, but the prehistoric flows contain the longest caves so far explored. Kazumura Cave has 11.7km of passage, nearly all along a single tube, separated by just one choke from a downflow extension which is nearly as long again. Many more lava caves await discovery, and the major explorations to date have been by visiting British and American cavers.

HONDURAS

A considerable area of limestone exists in the north-west region, but little serious exploration has been undertaken. The principal cave, and the only known deep one, is the Sumidero de Maigual (−420m) on the Montaña Santa Barbara; it consists of a rapid series of sporting shafts to a depth of 350m, followed by a long, large horizontal passage. Further east, caves have been explored in the Río Talgua area, but any entrances on the spectacular high karst of the Montañas de Colón remain hidden by dense jungle.

HONG KONG

There is very little limestone in either Hong Kong or the New Territories, but small caves up to 100m in length have been reported both around the coast and inland.

HUNGARY

While karst covers a total surface area of barely 1,350 sq km, it is renowned for its considerable richness and diversity of form. All the limestone has been explored with an intensity rare elsewhere in the world, due to major economic considerations involving mineral, water and energy resources.

Some fifteen hundred caves have been recorded, of which the finest occur in the vicinity of Aggtelek. Dolines, uvalas, depressions and large karren fields are characteristic of this lush district. Caves and potholes are numerous, and include the 25km-long Baradla cave and the adjacent 8,890m-long Beke Barlang. On the Also-hegy plateau, at the northern edge of the area, are many deep shafts (Vecsembukki Zsomboly, −245m). Slightly further south, in the higher Bukk mountains, caves are rich in archaeological, palaeontological and anthropological remains. Several other systems, such as the 2,940m-long and 240m-deep Istvan-lapa Barlang, are quite extensive, and many, known for their beauty, have been developed into show caves.

The rolling hills around Budapest house caves of a quite different character, many having been formed by the upward

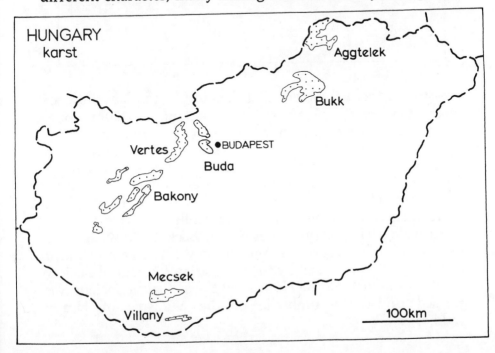

BARADLA BARLANG

Just to the north of Aggtelek, a range of low limestone hills, pitted with dolines, contains the great Baradla cave system. Its splendid main passage carries a stream in parts and is generally very well decorated; consequently some sections have been developed as show caves. There are some tributary passages, and one long arm of the system reaches under the international frontier into Czechoslovakia where 5km of passages are known as the Domica Jaskyna.

The cave has a long history of exploration, and by 1825 over 8km of passage had been mapped. The Baradla-Domica connection was made in 1926, and the system now extends to about 25km of passages, all within a depth of only 116m.

PLAN

pressure of thermal waters. This has resulted in labyrinths of great complexity (Mátyás-hegyi Barlang, 4,200m long, Ferenc-hegyi Barlang, 4km long). Few caves are found in the dolomite of the Vertes hills, but the nearby Bakon range has many small caves, together with the 170m-deep Alba Regia Barlang. The only other region of interest is that in the south, which, though caves do occur, is best known for the fine, bare karren fields and dolines of the Villany hills.

Speleological history can be traced back to 1549, when a record was first made of the Aggtelek caves. A commission for Speleology was founded by the Hungarian Geological Society in 1913, and in 1958 the Hungarian Speleological Society was formed under the umbrella of the Ministry of Heavy Industries.

ICELAND

The whole island is of volcanic origin, and there is no limestone. Lava tubes are common on many of the flat basalt plains. The major known caves are Surtshellir (1,970m long) and Stephanshellir (1,520m long) on the Hallmundarhraun plain in the west of the country, and Raufarholshellir (1,300m long) east of Reykjavik. The inhospitable climate makes these caves cold and bleak, but they can contain some splendid ice formations. On the Snaefells peninsula, the Bougarhellir cave (600m long) has some fine lava helictites, and there are caves recorded in many other parts of the island.

The ice caps of Iceland contain a number of glacier caves, the most remarkable of which have been created where warm geothermal springs rise beneath the ice. On the north side of Vatnajökull, the Kverkfjoll glacier has a magnificent underground river within the ice. The upper end of the system has been explored for 2,850m to a depth of 525m; most of the cave has a roof of ice with a floor cut down into volcanic tuff, and the descent includes a number of pitches.

Most cave exploration has been by visiting British, French, German and Spanish teams.

INDIA

This vast sub-continent, with its mighty northern mountain chain, contains disproportionately little in the way of interesting karst. Many areas still remain to be fully explored but it seems that there will be no caves on a world scale nor any karst features of outstanding significance.

Limestone is extensive throughout the Indian Himalaya, but karst of any important is known only in the Vale of Kashmir and in the Simla to Chakrata region. The former is best known because of its great vertical range and large resurgences, though it lacks any major caves, and the latter for its numerous potholes (Lower Swift Hole, −70m). The long Kaimur Hills, south of Varanasi, are a mixture of sandstone and limestone with many rock shelters and the 350m-long

Gupta Cave near Budoker. The rugged, relatively unexplored, jungle-clad hills west of Vishakhapatnam contain India's deepest system (Borra Guhalu, −86m), whilst the Erramala range by Kurnool has the longest (Belum Guhalu, 3,220m). Hampsi, in sandstone, and Cuddapah, in limestone, while frequently mentioned with regard to caves, are both minor regions with very small systems.

Perhaps the least explored area, but the one with the greatest potential is that of the Garo, Khasi and Jaintia hills within Assam. Numerous shakeholes, sinks, resurgences and caves have been reported, and the streamway of the Siju Cave, in the Garo hills, has been explored for 1,200m.

Recorded explorations commenced in the 1880s by various army officers, but it was not until a spate of foreign expeditions from 1970 that any detailed cave studies were made.

INDONESIA

Most of the great groups of islands, which comprise Indonesia, contain some outcrops of limestone, and due to the wet tropical environments nearly all of these are developed into cavernous karst. Except for the island of Java, most of Indonesia is thinly populated, and access is not easy. The cave potential of the country is enormous, and there is still much exciting karst unvisited by cavers. The major discoveries to date have been made by foreign expeditions, mainly British, though local cavers are starting to make their own explorations on Java.

The southern chain of islands, from Sumatra to beyond Timor, is formed by a very active chain of volcanoes, though these are mainly explosive and so have few lava caves. Limestones, originally formed as reefs, fringe most islands and are carved mainly into fine cone karst; the largest areas with most known caves are on Java. Sumatra has less limestone, but inland Gunung Ngalau Saribu translates as the 'mountains of a thousand caves', and many river sinks are known which may traverse the entire range in major cave systems; 4,000m of passages have been mapped in Gua Airhangat. The island of

Sumba is composed largely of limestone, and there are dozens of caves, though they are of limited depth; the longest yet mapped is near Lewa and has 2,500m of fine streamway. The karst of Timor remains unexplored for caves, though large river sinks have been reported in the interior.

The northern islands, from Borneo to New Guinea, have more complex geology, and their interiors are even less accessible. Kalimantan has a number of isolated limestone outcrops, some of which form dramatic plateaux fringed by cliffs; the longest known cave is Lubang Dunya (4,200m) in the Mangkalihar area on the east coast. The Maros area of southern Sulawesi, parts of central and eastern Halmahera, and some of the smaller Moluccan islands all contain landscapes of cone karst where sinking rivers and unexplored caves have been reported. Irian Jaya has perhaps the most exciting potential of all. Its western peninsulas contain large areas of rugged and spectacular limestone karst where sinking rivers and open cave entrances have been reported. The higher mountains in the eastern core have limestone rising to elevations of over 4,500m. North of Ngga Pulu (Carstenz) is a vast glaciokarst on the Kemabu Plateau, and there is a similar depth potential in the Mount Wilhelmina limestones. Cone karst, doline karst and glaciokarst of bare pavements are all found around Wamena in the Baleem valley. A number of caves are known, and the Baleem river sinks in a giant whirlpool and resurges from caves 2km away. The cave potential of the area remains exciting; only recently the Salukkan Kallang cave (8km long) was discovered in the Maros karst.

Java

Limestone karsts lie along both coasts, on either side of the central line of tall, active, volcanic mountains. Gunung Sewu, in the centre of the south coast, covers over 1,000 sq km and is not only the type area of cone karst but also the area currently having most known caves. In western Sewu over 250 caves are recorded, nearly all draining to the one coastal resurgence at Baron. Gua Jomblang (3,320m long) and Gua Bribin (3,900m) contain segments of the same major cave river, and the shaft systems of Buhputih, Ngepoh and Puleirang reach −200m.

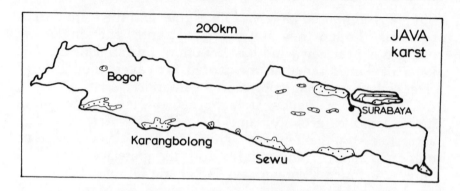

The finest cave in the area is Luwang Grubug (−161m, 2290m long), with its superb entrance shaft and powerful cave river. In eastern Sewu, Luweng Jaran (−158m, 11,250m long) is the longest cave in Indonesia, with a large main streamway and well-decorated high levels. Luwang Ombo (−230m, 2,900m long) has a fine 100m entrance shaft, and Gua Sodong contains 4,290m of streamways.

Elsewhere on Java the karst is generally less spectacular, but long caves have been found. In the Karangbolong karst, Gua Barat is a 3,300m-long resurgence streamway; Gua Si Wuluh, near Bogor, reaches −160m through chambers, explored by bird's nest collectors, who descended pitches of 40m on bamboo ladders while carrying paraffin flares for lighting. Long stream caves are known near Surabaya, and a large karst east of Sewu may hold considerable potential.

IRAN

The Zagros and the Elburz are two great ranges which form the basic structure of this large country. Limestone is extensive but in the areas so far investigated is rarely well developed into karst. Over the full 1,500km length of the Zagros, carbonates predominate. Many minor regions of karst are known, around Mahabad, Hamadan (Ghar Sar-Ab, 1km), Kuh i Garun (Ghar Garun, −112m), Mardesh and Bishapur. Only one area can be classed as being of major importance; this is the 150km-long

line of massifs situated to the north of Kermanshah between
Bisetun and Shapur. High-level plateaux are strewn with
depressions and large dolines, notably on the massifs of Parau,
Shahu and Ravansar; sinks, shafts and caves include Ghar

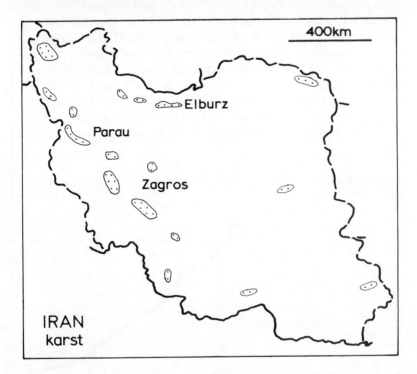

Parau and Ghar-I Cyrus (−265m). Large streams resurge
around the base of the mountains, but only around Ravansar
has it been possible to make any penetration (Grotte
Supérieure de Shabon Kale, 650m long).

The predominantly volcanic range of the Elburz neverthe-
less contains many areas of limestone. Numerous small caves
are known but nothing has yet been found of any significance.
Since 1971 Iran has been visited by British, French and Polish
speleologists with varying degrees of success.

GHAR PARAU

The spectacular massif of Kuh i Parau, just north of Kermanshah, contains Ghar Parau, Iran's deepest cave. It is situated just 300m below the 3,357m summit of the mountain in the karst of the south plateau. The cave is a strenuous proposition, being a continuously narrow, single passage, with twenty-six pitches up to 42m in depth. It was explored by British expeditions in 1971 and 1972, for 1,360m to a sump at a depth of 751m. The exact resurgence of the water is not known, but all the risings are around the base of the massif a further 1,000m down. None can be entered.

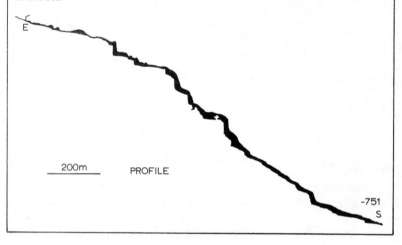

200m   PROFILE

-751
S

IRAQ

Thinly bedded outcrops of limestone are to be found in the eastern Badiet esh Sham, but it is not until the village of Al Hadithah that karst features are noticeable. These consist of sinks (Salmon Rose Sink, 50-70m in diameter, 45m deep), ponors, dolines and caves (Fasaya, 1km long). Blind cave fish have been reported.

In southern Kurdistan, in the east of the country, access has always been severely restricted. However, many of the mountains north of the Dukan Dam and within the Great Zab basin are formed of limestone. Some have developed a jagged alpine character, while others, formed of massive horizontal beds, have become undulating plateaux terminating in great cliffs; in the Rawandiz region these are cut by deep ravines. Caves exist in many areas. Lapies and depressions are common and reported as being very fine above Dihok.

Minimal cave research has been carried out, by such universities as Mosul, and this has been in the pursuit of water resources.

IRELAND

Over half of the area of Ireland (excluding the province of Ulster) has limestone as bedrock, but the central lowland has an extensive peat cover and minimal karst. Upland limestone around this central plain does, however, have some fine karst with caves in various isolated areas.

The most important karst in Ireland is that of Co. Clare. Its eastern half forms the Burren, with some of the finest pavements in Europe; large shallow depressions drain underground but there are few caves other than the Fergus River Cave, 2,300m long near a main resurgence. The western half of Co. Clare is a fine cave area with sinkhole systems all around the shale hills of Slieve Elva and Poulacapple. Most caves are clean washed, gently graded canyon streamways. They include Pollnagollum, with over 12km of passages, and the long parallel Cullaun caves (of which Cullaun 5 is 4,800m long). Towards the coast the Coolagh River Cave (4,100m) is a classic, flood-prone system; the Doolin cave system has 10,500m of stream passages and provides an unusual dry through route beneath the flowing River Aille; nearby Pol-an-Ionain is only short but is famous for its single 7m-long stalactite.

Other karst areas in the south are restricted to thin outcrops of limestone. At Michelstown, in Co. Tipperary, the New Cave has a maze 2,300m long, and there are other, shorter fossil caves in the area around Cork. In Co. Kerry the most important area is at Castleisland, where the recently explored Crag Cave is now 3,810m long.

Further north, there is an extensive lowland karst in Co. Galway and Mayo. Lough Mask drains underground to Lough Corrib, without any known accessible cave, though there are fine pavements in the area. A feeder to Lough Mask is the Aille River, which goes underground through a major and very aqueous cave of the same name (over 1,900m long). The River Abbert drains into Corrib, after partly flowing through Ballyglunin Cave (a 1,500m warren of shallow passages).

Within the counties of Sligo, Leitrim and Cavan are some fine limestone plateaux, though the best caves are across the border in Fermanagh. Pavements, sinkholes and marginal cliffs identify the karst areas, though most caves are short,

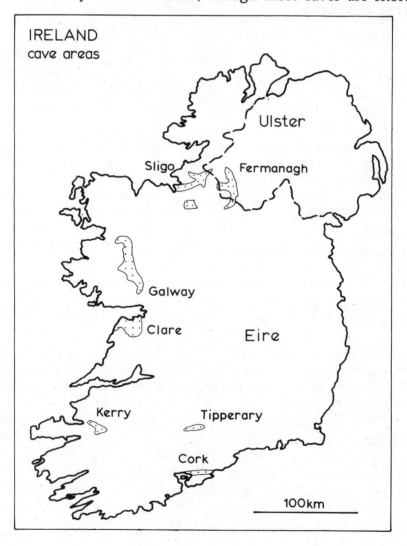

IRELAND
cave areas

Ulster

Sligo   Fermanagh

Galway

Clare   Eire

Kerry   Tipperary

Cork

100km

with limited streamways. Polliska (−128m) is a stream cave with a staircase of shafts, and Carrowmore Caverns is a complex of rifts reaching to −140m. The mighty River Shannon has a karst resurgence at its head, though the only explored feeder cave is Pollahune with its 1,500m of streamway.

Most cave exploration has been by visiting British cavers.

IRELAND, NORTHERN

The politically isolated northern province of Ireland, often referred to as Ulster, contains one major region of karst and caves in the county of Fermanagh, right against the border with Ireland. The limestone stretches along the hills south-west of Enniskillen, with Tullybrack and Cuilcagh separated by the Belcoo valley.

There are dozens of sinkholes where streams draining off Tullybrack meet the limestone on its north-eastern slopes. The most important cave system is Reyfad Pot (6,500m long); its broken 100m entrance shaft drops into large fossil passages, though a lower streamway reaches a depth of 179m. It is also entered by the Pollnacrom shaft, and there is considerable potential towards the Carrickbeg Rising. Nearby, Noon's Hole is a fine 80m shaft into the upper end of the splendid streamways of Arch Cave (4,500m long). Pollaraftra is another long stream cave, and there are more shafts on the mountain, many with potential for further discoveries.

The fine limestone karst around Cuilcagh has been well explored only on its northern side. The Marble Arch cave system (6,500m long) drains into the head of the beautiful Claddagh Glen; it has a number of large stream passages, mostly entered from the resurgence or from large collapse dolines between there and the sinks. Just to the east, Prod's Pot has a series of shafts into a fine streamway which can be followed through sumps into the long Cascades Cave and out to the resurgence; sink to rising is 3km, and the system is 4,800m long. The Tullyhona Rising has a streamway cave 2km long behind it, and there are many potholes with considerable potential on the limestone hills to the south.

Most of the major explorations have been by visiting British cavers, though the Irish cavers have done more work recently and are now very active.

ISRAEL

Two limestone ranges dominate the northern and central regions of Israel. The Galilee Hills form a lower extension of the Jebel Liban. Karst features are to be found throughout, but the most notable areas are around Har Meron and Har Kena'an. Here lapies has developed on the now dry plateaux, with many small caves on the intervening slopes. These include the country's deepest system, Hotat (−147m). South of the Plain of Esdraelon lies the Judean Platform, where enclosed basins, depressions and sinks are common. Small caves, rich in palaeolithic material, are known on Mount Carmel, and similar systems exist to the north of Jerusalem. In the maze of limestone canyons around Qumran are the caves which once housed the Dead Sea Scrolls.

At the southern end of the Dead Sea, on the long, low Mount Sedom, there is a very pronounced salt karst which includes funnels, lapies, shafts and caves. The latter include the longest and deepest known in the world in salt; Malham Cave has 3,100m of passages and is 130m deep. Nearby there are three shafts over 50m deep and also Sedom Cave (1,060m). Further south, in the Negev, limestones do outcrop but no karst features are known.

ITALY

Limestones occur extensively throughout both the Alpine and peninsular regions. In most areas karst phenomena are varied and well developed, with particularly spectacular landscapes on Monte Canin, on the Apuan Alps and on the Gargano and Murge regions of the peninsula. More than ten thousand caves have been recorded, many of world class, reaching to over 1,000m deep and over 40km in length. The potential for new exploration is still considerable, and these statistics should soon be surpassed.

The history of Italian caving dates back to early research in the karst of Trieste, by Linden in 1839, and the formation of a

Commissione Grotte in 1883. Interest in caving rapidly developed, so that by the early 1900s there were many local groups in the northern half of the country, and in 1929 a national Istituto Italiano di Speleologia was organized. A steady flow of discoveries has been maintained ever since and has developed into an explosion of activity over the past twenty years.

Alpine Italy

Limestone is extremely common over much of the alpine region but the finest karst and deepest caves occur at its western and eastern extremities – on the Marguareis and Canin massifs respectively. The former is the karstic climax of the long combined chains of the Maritime and the Ligurian Alps. This extends along the southern French/Italian border and then towards Genoa. Well over a thousand caves are known, but around Monte Marguareis no fewer than six extend to a depth of 500m or more. These include the Complesso di Piaggia Bella, Abisso Cappa (−706m) and Abisso Straldi (−614m). The surface karst is an exciting one of barren lapies, dry valleys, rugged dolines and deep shafts.

A wide band of limestone expands from the shores of Lake Garda eastwards to the Yugoslavian border. Caves are to be found everywhere, but important systems have so far been found in only a few areas. The combined western districts of Varese and Como are primarily noted for their many short systems of great beauty, as well as the much greater finds on and around Monte Palonzone. These include the recently explored Abisso di Monte Bul (−557m) and the Grotta Guglielmo (−452m).

Continuing eastwards, the rugged Bergamo Alps, with their scenery of peaks and crags, have much lapies, many small caves and also the 428m-deep Buco del Castello. The areas of Asiago and Lessini are almost connected, the former being an undulating plateau averaging 1,400m in height. Dolines and great pits, up to 50m in diameter and depth, are a characteristic of both regions. Caves are mainly horizontal, with the longest system, the complex Bus de la Rana, on Monte di Malo, extending for over 21km. The rugged mountains and richly wooded valleys of Lessini contain many karstic gems. These

COMPLESSO DI PIAGGIA BELLA

The seven entrances of this very popular Italian system are to be found on the western flanks of the Marguareis massif, not far from the French border. The original entrance, Carsena di Piaggia Bella (altitude 2,163m), was first investigated in 1944, and the terminal syphon was reached with relative ease in 1958. The main difficulty of exploration was the boulder mazes in the large entrance chambers, but beyond them the main streamway provides a gentle descent.

The other six entrances are the Abisso Jean-Noir (altitude 2201m), Abisso Caracas (altitude 2297m), Abisso Indiano (altitude 2195m), Buco delle Radio (altitude 2,160m), Abisso Solai (altitude 2,037m) and Abisso S2 (altitude 2,357m). These all join together at varying depths (to −554m) and have formed a 24km complex of passageways 761m deep. Whilst the Piaggia Bella cannot lay claim to any major proportions on a world scale, it is perhaps the most frequently visited of any major Italian cave. Its popularity was given an added impetus in 1982 with the discovery of the Abisso S2 and its connection with the main system, thereby creating a classic traverse over 6km long. Unlike the original entrance, the S2 and the other inlets contain fine, steeply descending streamways and shaft sequences.

The water within the cave resurges at Foca (altitude 1,180m) but little progress in this direction is expected.

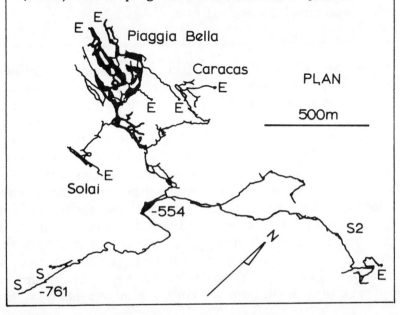

SPLUGA DELLA PRETA

This historic Italian cave is found at an altitude of 1,475m on the gentle southern slopes of the Monte Lessini massif not far from the village of Sant'Anna di Alfredo. Its great 131m entrance shaft, situated in the floor of a doline, has been known since 1825, but it was not until 1925 that any attempt was made at its descent. By 1927 exploration of three deep pitches reached a stated −637m, and it immediately claimed the world depth record, a position it held for many years. Much later, new exploration and a resurvey revealed that this depth was in fact only −376m. Around 1960 serious interest again commenced in the system, and a series of narrow canyons and fine shafts were descended. In 1982 the current depth of −985m was reached, when a traverse over a shaft at the −840m level revealed a final extension.

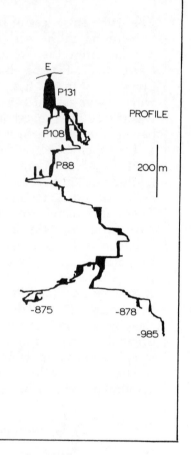

take the form of small but perfect poljes, fields of dolines, superb examples of lapies and the famous Spluga della Preta.

The eastern region of Italy, which protrudes into Yugoslavia by the Gulf of Trieste, contains the much-studied and recorded Italian karst. This consists of two plateaux named after the nearby towns of Monfalcone and Trieste. Both areas are dominated by spectacular doline karst with broken and chaotic lapies. Faults have influenced cave development, and none of the 230 systems reaches 1km in length or has so far exceeded the 331m depth of the Grotta di Trebiciano. The great challenge of the area is the tracing of the mysterious River Timavo; it commences its journey at Skocjanske Jama in Yugoslavia, where, after 2km, it disappears into a syphon. Twenty kilometres later, at the bottom of Trebiciano, it can be followed again for a brief 200m. The water then resurges from the eighteen springs at San Giovanni di Duino, nearly 50km from the place where it first started its underground journey.

Monte Canin

This mighty massif sits astride the Italian and Yugoslavian border to the north-east of Udine. The barren and inhospitable plateaux to be found at the 1,700m to 2,000m levels are considered to have some of the finest and most spectacular examples of lapies, dolines, shafts and fissures within Italy.

ABISSO MICHELE GORTANI

High on Monte Canin, four entrances descend through some of Italy's most spectacular lapies. These then connect at the 296m, 514m and 550m levels to form an 8,325m complex of passageways descending to an arduous depth of −920m. The original entrance at an altitude of 1,900m was first discovered and explored to a depth of −240m in 1965, and in 1969 the bottom was reached, at −892m. The following year another entrance, the Alto Ingresso, was found some 28m higher, and this gives the current depth. In 1975 two further systems, the Abisso U2 and Abisso A12, both at slightly lower altitudes, were connected to give the cave its greater length. The water from the terminal lake resurges at the Fontanon di Goriuda (altitude 868m). The Gortani has recently been connected to the Abisso Davanzo, making a total system length of 13km, though the depth remains unchanged at 920m.

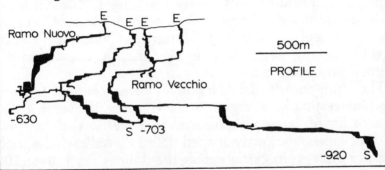

Most cave systems are entered at this altitude, and currently eight have been found to reach depths of more than 500m. The deepest of these caves is the Abisso Michele Gortani, followed by the Abisso Emilio Comici (−774m), the Sistema S20-S31-FDZ2 (−760m) and Abisso Paola Fonda (−750m).

All the deep systems are characterized by long, narrow meanders punctuated by cold, well-watered shafts, and their exploration is therefore not easy. While there are several risings for the massif, the principal one is the Fontanon di Goriuda (altitude 868m). Speleological investigations on the mountains did not commence until the early 1960s, and the number of deep discoveries still being made each year suggests the full potential has not yet been realized.

Peninsular Italy

The Apennine range is the backbone of peninsular Italy, stretching for almost the full thousand kilometres of its length. Limestone plays an important part in the make-up of these mountains, and many major areas of karst exist, not all of which have been fully investigated.

In the north-west are the Apuan Alps, a range of such great importance that it alone can ensure Italy's place among the world's greatest caving nations. Close by, just to the south of Bologna, is an interesting region of gypsum karst. The rolling landscape is marked by scattered dolines and small blind valleys, many of the latter terminating in caves. Some of these reach considerable proportions, such as the connected system of Grotta della Spipola and Buca dell'Acqua Fredda, 7km long.

Along the borders of the Marche and Umbria, not far from Perugia, is the Monte Cucco area, a region of rather uninspiring rounded hills barely reaching 1,700m in height. Two deep systems exist here, the Grotta di Monte Cucco, and also the steeply descending Grotta del Chiocchio (−514m). Slightly to the east, towards Ancona, is the 21km long Fiume-Vento complex.

The Promontorio del Gargano is often claimed to be the most interesting karst region in the country. It rises to 1,000m in a series of large plateaux, many covered with extensive fields of lapies, variously shaped dolines, swallowholes, poljes and dry valleys. In certain areas the dolines reach over 100m across, and one, the Dolina Pozzatina, measures 675m by 440m by 130m deep. Caves are also common but none reaches major proportions.

Just south of Gargano is Le Murge, a large, low-lying plateau, featuring numerous dolines, sinkholes and partly enclosed depressions. Of particular note are the *puli*, large

GROTTA DI MONTE CUCCO

The great, rounded, 1,566m-high mountain of Monte Cucco is situated in the province of Umbria close to the town of Costaccioro. It has been found to be spectacularly hollow and houses the 922m-deep and 26,130m-long Grotta di Monte Cucco. Large, gently dipping passageways descend into the hillside from two entrances at 1,390m and 1,509m. These galleries separate into several levels and lead to enormous chambers and even greater shafts – the final pitches to the bottom are 171m, 116m and 109m!

The lower entrance was first entered in 1890, but it was not until 1967 that any extensive exploration was carried out. By 1973 the cave had reached −791m, +131m, and the following year this upper passageway was forced to the surface just 57m from the mountain summit. The little water in the cave resurges at Sarco, some 2.5km distant, at the 584m level.

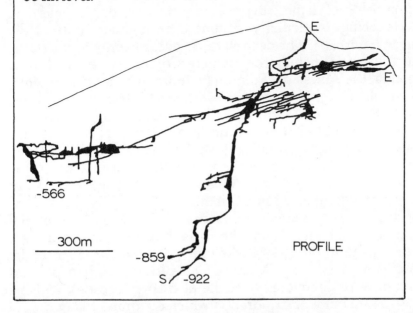

-566

300m

-859

S -922

PROFILE

open shafts, one of which, the Pulo di Altamura, is 727m across and 96m deep. With the exception of the spectacularly decorated show cave of Grotta di Castellana, caves are few and small.

The remaining areas of the south are incompletely explored but deep systems do occur on the Matese massif, which sits astride the Caserta and Campobasso provinces, and on the Alburno and Cilento massifs of Salerno. In the former region lies the Pozzo della Neve (5,000m, −895m), and in the latter four caves exceeding 300m in depth have been found in recent years. The most southerly Italian karst which contains any major systems is that of Monte Pollino in northern Calabria. This contains the very sporting Abisso di Bifurto, whose staircase of clean shafts reaches a depth of 683m.

Apuan Alps

The limestone mountains of the Apuan Alps, east of the town of Massa, cover an area of approximately 250 sq km. Originally famous for the fine marbles of Carrara and Massa, they are now equally renowned for the great collection of deep caves. The maximum elevation is moderate, with the highest point being 1,945m on Monte Pisanina, and the average is around 1,100m. The whole range exhibits karst features, with major fields of lapies, depressions, dolines, dry valleys and shafts being particularly common throughout a decidedly barren landscape. The deepest caves occur mainly in the centre and south and include the Fighiera Corchia system in Monte Corchia, the Abisso del Fulmini (−760m) on Monte Altissimo, Abisso dei Draghi Volanti (−870m) on Monte Sumbra, Abisso Paolo Roversi (−755m) on Monte Tambura, and Abisso Oriano Coltelli (−730m) on the Alto di Sella. The area is also noted for its profusion of deep shafts, many directly from the surface, and these include the Pozzo Mandini (310m), in the Abisso Roversi, and the Abisso Revel (300m) and Pozzo Giovanni (226m) on Monte Corchia.

Cave investigations commenced in 1926 with a partial exploration of the Tana che Urla but did not really get underway until the 1930s, when the potential of the Antro del Corchia was realized. Many other systems with depths to −400m were discovered over the next three decades, and the 1970s saw a burst of exploration which continues today.

COMPLESSO FIGHIERA FAROLFI CORCHIA

This system, the deepest and greatest of Italian caves, is situated beneath the rugged slopes of the 1,677m high Monte Corchia in the Apuan Alps. It was won in 1983 with the joining of three major caves – the Antro del Corchia, the Abisso Claudo Fighiera and the Abisso Radolfo Farolfi. The result is a 45km complex of galleries, great chambers and shafts on a multitude of levels, with a total vertical range of 1,210m.

The Corchia is often equated with the Gouffre Berger as one of the world's most sporting and enjoyable caves. It has four entrances leading into a complex of fossil high levels, and a fine lower streamway with many cascades to the terminal sump. The Fighiera has just one entrance, at an altitude of 1,640m, and the major explorations started in 1977 with the discovery of a long descending passage to shafts ending in a choke over 500m down. Subsequent explorations revealed high levels and branches of increasing complexity, reaching towards the fossil level in the Corchia. The Farolfi cave was first found in 1980 and was connected to the Fighiera passage the next year.

The water from the system resurges at an altitude of 350m by the village of Cardosa. It is unlikely that the cave will be extended in this direction; nor is it likely to be extended upwards as it is already only 37m from the summit of Monte Corchia, though there are considerable possibilities for further horizontal passage. Most exploration has been by Italian cavers, with the exception of a major extension in the Corchia, found by British cavers in 1968.

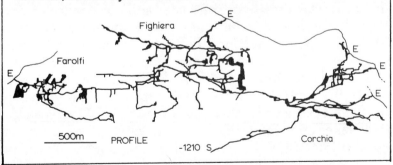

Sicily

This fascinating island, off the 'foot' of Italy, has caves and karst in three different rocks. The limestones of north and south Traponi, Palermo, Syracuse and Ragosa all contain dolines, depressions, sinks and dry valleys. Notable caves are to be found only on Monte Pellegrino (Palermo) and include the 2,000m-long Grotta Addaura III and the 202m-deep Abisso della Pietra Selvaggia. Gypsum karst is extensive in both central Traponi and around the town of Casteltermini. In the east of the island, Mount Etna has many lava caves in its basalt flows; the Grotta del Lamponi is a fine large tube 1,700m long.

IVORY COAST

There are no areas of karst, but tectonic fissures are known to reach a depth of 110m in the Man massif.

JAMAICA

Jamaica was perhaps the first 'caver's paradise' to be found — with its tropical climate, warm water and abundance of caves. Over half the island is limestone, with the major cavernous areas in the central section. The cone karst of the Cockpit Country and Dry Harbour Mountains lies north of a central anticlinal ridge of impermeable rock centred on the village of Troy; to the south the limestone is repeated. Many fine river caves flank this central inlier; Quashies River Cave is a classic with its lakes and waterfalls, but Still Waters River Cave is the longest in Jamaica, with 3,600m of mainly aqueous horizontal passage. Away from the central inlier, there are extensive fossil caves and scattered deep shafts. Of the former, Jackson's Bay Cave is notable for its decorations, and of the latter Morgan's Pond Hole is the deepest, at 186m. Many of the major discoveries have been made by visiting British cavers, and there is still plenty of potential in the forested karst, though the Jamaican Caving Club have recently been more active.

JAN MAYEN

Limestone does not exist on the Arctic island, and no volcanic caves have been noted on the lava fields.

JAPAN

The great archipelago of mountainous islands which form this country is endowed with a multitude of contrasting karst regions. None of these areas is particularly large, but they usually occur as plateaux or blocks of thick limestone cradled against a background of volcanic hills. Caves and potholes are common, but because of the unusually high number of karst districts there is still considerable potential for the discovery of many more small to medium-sized systems.

On the island of Hokkaido the karst is little developed underground, and no caves of any note have so far been recorded. Further south, in the extensive Iwate karst of Honshu island, several of the country's longest and most interesting systems are to be found. These include the 8km-long Akka do, the 3,200m Uchimagi do and Ryusen do, with its reputed 100m-deep lake. Close to the west coast is the Niigata ken karst where all the deepest caves are situated (Byakuren-do, −450m; Omi-Senri do, −402m). The long sweep of outcrops between Fukushima and Tashaku-dai karsts so far contain no systems of great importance, but at the south-west tip of Honshu is Japan's most renowed and intensively researched region, Akiyoshi-dai. On this spectacular plateau karst of lapies, dolines, depressions and poljes there, there are bamboo groves, paddy fields and more than 250 caves. The longest of these is the show cave of Akiyoshi-do, with its 2km of passages containing a fine resurgence streamway and many spectacular decorations, and nearby lies the 145m-deep Nishiyama-no-Tateana.

The Ohnogahara karst of Shikoku consists of scrub-covered lapies and dolines, with horizontal caves up to 2,400m in length (Ryugo-do). Kyushu island has two main karsts; Hirao-dai is a barren plateau with numerous caves and occasional deep potholes (Gomagara do, −260m), and Kumamoto-dai contains many lesser systems.

The southerly Ryukyu group of islands is very rich in limestone and karst, and the 60km-long Okinawa island is riddled with over two hundred caves. These include the part show cave of Gyokusan-do (5,000m) and Todaroki (2km+).

Lava caves abound throughout the islands of O Shima, Hachyo jima, Fukue and Daikon, and also around Mount Fuji (Honshu) and Mount Aso (Kyushu). Few caves reach 400m in length but one tube on Fukue, now much collapsed, was originally 1,400m long.

JORDAN

Very thinly bedded limestone extends along much of the northern half of the Rift Valley, and small caves and minor karst features have been noted in the Irbid and Karak regions. Further east, the limestone has been covered by sand, but around Petra there are many small sandstone caves.

KENYA

The caves of this exciting East African country are almost equally divided between lava and limestone. In both cases further potential is considerable. Most of the exploration has been by the local Cave Exploration Group since its formation in 1972.

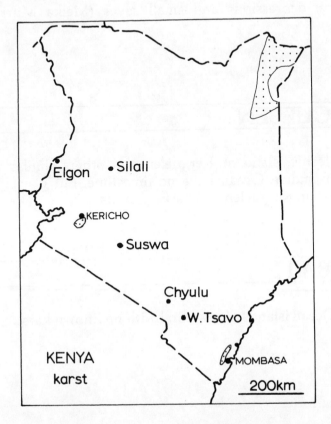

On Mount Suswa over fifty entrances to lava caves have been found which lead to a total of more than 8km of explored passages – the longest continuous length being 750m. Between the 2,000m and 4,000m level on Mount Elgon many more caves exist, including the often-reported Elephant Caves (entered by large game, including elephants, in search of mineral salts). The Chyulu Hills, to the south, contain almost as many lava caves as Mount Suswa and include one of the world's longest, the 12,000m Leviathan Cave (there are several entrances). More, but lesser, systems are known in the West Tsavo National Park, on Mount Silali, and are reported to occur on the volcanoes around Lake Rudolf.

Limestone karst has been extensively explored just inland from Mombasa where there are shafts, dolines and sinkholes, together with many cave systems up to 500m long and 50m deep. More small, but much wetter, caves are known in the hills between Kericho and Satik. In the very north-eastern corner of the country is a vast, little-explored limestone plateau with many dolines up to 50m in diameter (Arda Gabiche), depressions and small caves (Melka Murri Pot, −50m).

KERGUELEN

The volcanic island of Kerguelen and others similar in the southern Indian Ocean have no limestone, but glacier caves may exist on Kerguelen and Heard islands.

KIRIBATI

Pacific Ocean islands of low coral with no known karst.

KOREA, NORTH

Little information is available on this potentially exciting country, but both limestone and karst are known to be extensive in the mountains to the east of Pyongyang, where one cave, Dongryong-gul, exceeds 1,500m in length.

KOREA, SOUTH

Extensive areas of limestone and karst are to be found throughout the central Sobaek mountains range and eastwards to the T'aeback hills and coast. Much of this area has been only partially explored speleologically but already over a thousand caves are known, including a dozen over 1km in length. Most of the deeper and longer ones occur in the south-west of Gongwean province and include Hwansun-gul (8,700m) in Samcheog-gun, Chondang-gul (6,000m) in Samcheog-gun and the 181m-deep Namgamdock-gul in Yeongweol-gun. Other caves are to be found further south and westwards, but these are so far relatively small.

The island of Cheju do, which consists of one major volcano, Halla san, and is surrounded by some 360 minor cones, is very rich in lava caves. At least eight systems are known over 1km in length, with the longest being Bilremot-gul, 12,400m.

There is a very strong nucleus of cavers in Seoul who frequently make joint expeditions with the Japanese, with whom most of the work is published.

KUWAIT

Of no speleological interest.

LAOS

Massive, barren and desolate karst ranges are to be found extensively throughout almost all the mountains north of latitude 17 degrees. The major regions are situated north of Thakhek (Khammouane province) and then eastwards as far as the Vietnamese border, and to the north of the city of Luang Prabang. Due to political problems, almost no serious speleological work has been carried out, but it is known that long horizontal caves exist within the two regions mentioned (Se Bang-Fai, 4,200m and Nam Hin-Boun, 4,000m, both in Khammouane).

LEBANON

There are two major mountain chains within the country, Jebel esh Sharqi (Anti Liban) and Jebel Liban, both composed predominantly of Jurassic limestone. The former lies along the

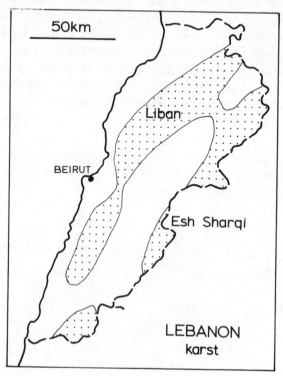

50km

Liban

BEIRUT

Esh Sharqi

LEBANON
karst

border with Syria and is basically a simple anticline 100km long with jagged peaks and barren slopes. Caves, dolines and shafts do exist but little serious exploration has been possible due to political tensions. The much larger Jebel Liban forms an imposing central backbone of the country reaching 3,086m in altitude. The highest peaks are formed of very pure Cretaceous limestone and exhibit extensive karst features.

FAOUAR DARA

The Lebanon's major system, situated at an altitude of 1,598m, is entered on the slopes of Jabal Zaarour, north-west of Zahle. A small karst valley leads down to the cave, which exemplifies all that is good about caving. Large well-washed passages dominate, with deep potholes, cascades, some twenty-two pitches and beautiful formations. Faouar Dara has been known since an initial visit, in 1955, reached the top of a 119m shaft. Expeditions in 1959-62 reached the terminal syphon, at −622m. Subsequent visits extended other passages, resulting in a total length of around 3,500m.

The water resurges some 20km away as a spring just 30m above sea-level at Faouar Antelias. The underground river can be reached through the nearby system of Magharet El Kassarat and followed along a superb streamway for almost a kilometre to a syphon (total length, 2,400m).

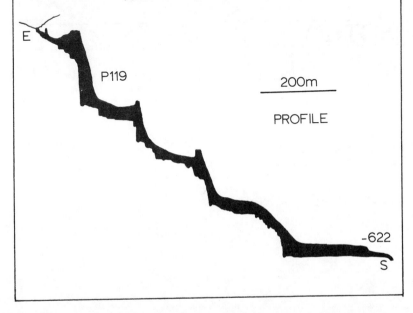

E

P119

200m

PROFILE

−622

S

Well over a thousand caves are known throughout the range's 160km length, but the major region is to be found within the area bounded by the towns of Beirut, Zahle, Amioune and Les Cédres. Here, within a rugged terrain of dolines, depressions, dry valleys and occasional lapies, are to be found such systems as Faouar Dara, the almost unbelievably beautiful river cave of Jietta (Nahr el Kelb, 8,330m long, +141m) and the complex Magharet el Roueiss (Aaquara, 5,066m long). The nearby karst plateau of Jebel Kesrouane contains Houet Badaouiye (−202m, with a 164m entrance shaft). Further potential is considerable.

LEEWARD ISLANDS

Antigua, Anguilla and Barbuda are the only islands in this group to be composed principally of carbonate rocks. No caves are known. Of the remaining volcanic islands, only Guadeloupe and in particular Grande Terre, Desirade and Marie Galante contain any caves, with the major system in the latter island – Grand Trou à Diable (−76m, 500m long).

LESOTHO

Of no speleological interest.

LIBERIA

Of no speleological interest.

LIBYA

Karst terrain is to be found on Jabal al Akhdar to the east of Benghazi. The range is essentially a limestone plateau faulted into three blocks. The most northerly of these plunges directly into the sea, and it is at this level that caves are situated. These include Bukarna (101m deep and 2,250m long – 78m of this depth is flooded) and Merisi, which has 1,673m of drowned passage out of a total of 2,100m.

South of Tripoli, between Al Ghanam and Yafrin, is a range of gypsum hills. Karst, with sinks, dolines, springs and shafts is extensive, and caves are known with a total of 7km of passages, including Umm al Masabih (3,593m long).

LIECHTENSTEIN

There is nothing of interest to speleologists within this small country, though it does have an active caving group.

LINE ISLANDS

A chain of low coral islands in the Pacific Ocean, with no known karst.

LUXEMBOURG

The Jurassic limestones of the south contain several caves. Notable amongst these are the 4km-long Caverne de Moestroffe and the 40m-deep Grotte Saint-Barbe.

MACAO

Of no speleological interest.

MADAGASCAR

The island of Madagascar, unique in so many aspects of its natural history, contains vast areas of little explored limestone. Over 33,000 sq km of karst have been recorded, all in the island's west. The surface morphology is mainly one of rugged plateaux edged by impressive escarpments. As the climate varies from semi-arid in the south to monsoonal in the north, the diversity of karst landscapes is considerable. Most cave explorations have been by French expeditions.

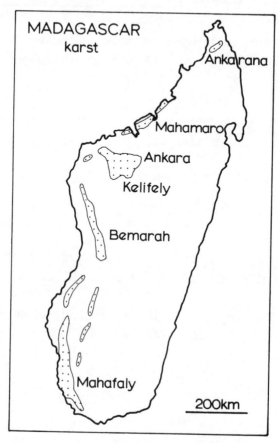

MADAGASCAR
karst

Ankarana

Mahamaro

Ankara

Kelifely

Bemarah

Mahafaly

200km

Five major karst regions can be identified, and the best known is the Ankarana/Andrafiamena massif in the north. This is a dramatic karst, fringed by vertical cliffs and carved into deep dolines with fault-guided canyons and a chaotic local relief of pinnacle karst (known locally as *tsingy*). The Ankarana massif contains over 80km of mapped cave, nearly all of it massive horizontal tunnel; the main river passage can be followed for over 5km in galleries nearly 30m wide. Each of the caves of Antsatrabonko, Andrafiabe, Anjohi Milaintely and Ambatoharanana contains around 10km of mapped passages. All the caves, and some of the dolines they connect to, contain spectacular suits of fauna and flora.

Tropical karst characterizes the 700 sq km Mahamaro region where many caves are known, including the 5,330m-long Andranoboko systems. The Kelifely/Ankara block exceeds 8,000 sq km in area and is little explored. It is pitted with dolines and depressions, and some spectacularly large shaft entrances have been seen from the air. Most southerly of the wet karst regions is the Bemarah plateau, rich in small caves, dolines and pinnacles. Mahafaly is the largest karst of all, covering 9,000 sq km. It is very dry, with only moderate relief, but it is pitted by large dolines and does contain the two deepest caves yet explored in the country (Gouffre de Tolikisy, −160m, and Aven de Laraboro, −115m).

MADEIRA

No limestone is to be found on this volcanic island but several small lava caves are known (Grutas de Cavalum, 83m long).

MALAWI

Minor karst features have been noted close to Zomba, but there are no records of any caves.

MALAYA

Very pure limestones and dolomites occur extensively throughout central and northern peninsular Malaya, now part of Malaysia. Karst areas stand out from the surrounding terrain due to the characteristic marginal rock cliffs, and also the distinctive flora (semi-drought vegetation due to the rapid percolation, as opposed to normal rain forest). Almost all contain extensive karren, dolines and dry valleys, and numerous caves; tower karst occurs in Kedah and Perlis.

Serious cave explorations have been limited. The largest known caves lie in the northern Perlis district; the Wang Tangaa depression can be entered only by a walkway through the 300m-long river cave which drains it, and an inlet cave is known to be over 2km long. All the limestone hills of the Kinta Valley contain caves, where Gua Tempurong is a 1,500m-long, large river cave. Many cave entrances are favoured sites for temples and shrines. Near Kuala Lumpur, the famous Batu Caves include the short Temple Cave and also the massive

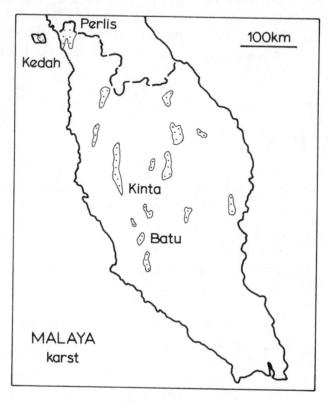

passages of the 1,500m-long Dark Cave. Many caves are threatened by limestone quarrying, and others have been modified or excavated by alluvial tin-mining.

MALDIVE ISLANDS

The low coral atolls of the Maldives, Lakshadweep and Chagos in the Indian Ocean have no known karst.

MALI

Very broken limestone occurs in various regions, but no karst is known.

MALTA

The five islands which make up this Tertiary limestone archipelago contain many small caves. These are to be found along the craggy coastline and are characterized by low entrances leading into one or two sea-filled chambers. Inland caves are to be found only on the main island of Malta and include the palaeontologically important Ghar Dalan (240m long) and the complex Ghar Hassan (300m long).

MARIANA ISLANDS

In the western Pacific Ocean, a chain of islands each with volcanic cores and coral limestone fringes. A recent American expedition found one large cave on Saipan and many smaller ones on Guam, Rota and Tinian. Pagan also has small lava caves.

MARQUESAS ISLANDS

These Pacific islands are of volcanic origin and possess no limestone. The rugged coastline is renowned for its many eroded sea caves, while inland several small lava caves are on record.

MARSHALL ISLANDS

Low coral islands in the Pacific Ocean with no known karst.

MAURITANIA

Limestone is commonly interbedded with sandstone, but apart from springs no karst features are known.

MAURITIUS

Very old volcanic islands with no speleological interest.

MEXICO

Though well known for its northern deserts, the landscape of Mexico is dominated by forested mountain ranges with high rainfall. The mountains are on a massive scale and include huge areas of limestone. Consequently Mexico is one of the world's major cave regions. There are already fourteen known cave systems with over 5km of passage, and seventeen caves reach depths greater than 500m.

The Sierra Madre Oriental, parallel to Mexico's east coast, is formed largely of limestone and constitutes the prime cave region. Further south the limestone continues into the geologically more complex southern mountain ranges which again have spectacular karst regions. Across to the west, the Sierra Madre Occidental and the central plateaux are dominated by volcanic rocks, though there are isolated blocks of limestone with karst. The Cerro Grande mountains in Jalisco contain the stream cave of Resumidero de Toxin (2,900m) and the nearby Resumidero del Pozo Blanco (241m deep with a 233m shaft). Further scattered caves occur as far north as Sonora and out in parts of Baja California.

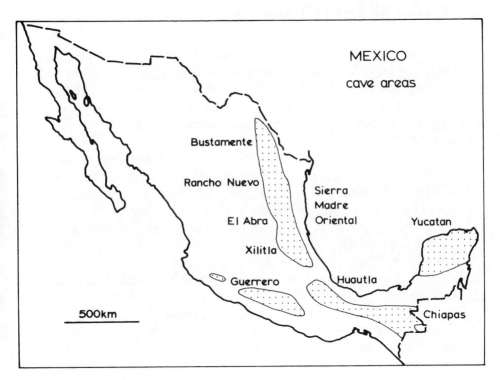

MEXICO

cave areas

Bustamente

Rancho Nuevo

El Abra

Xilitla

Guerrero

Sierra Madre Oriental

Yucatan

Huautla

Chiapas

500km

A distinctive and famous karst forms the Yucatan peninsula. There is little relief in this huge limestone lowland, which is characterized by flooded sinkholes and the flooded shafts known as cenotes. The ancient sacrificial well at Chichen Itza is one of the better-known cenotes, with vertical rock walls dropping into the depths. Dry caves in the Yucatan are thinly spread; Actun de Kaua has a maze of large passages extending for 6,700m, and the next longest cave is Actun Xpukil (3,300m). Some phreatic passages have also been explored by divers in a few of the cenotes.

Many of Mexico's large dry caves have been known to the local people for centuries, but serious exploration started only around 1965 when US cavers realized the vast potential lying south of their border. Most major discoveries have been by the Americans, though Mexican caving groups, the Canadians and various visiting European teams have all made their contributions. The potential for future exploration is still vast, and there is no doubt that Mexico has a generous share of the world's long and deep cave systems.

Sierra Madre Oriental

From north to south the Sierra Madre Oriental is a spectacular range of high limestone plateaux and escarpments separated by deep canyons. It is a rugged karst, with many huge dolines and deep shafts, and it covers a range of climatic zones giving vegetation cover varying from semi-arid scrub to thick rainforest.

In the north, the Bustamente area has a number of caves including the Gruta del Palmito with its massive, well-decorated chamber, and Gruta del Precipicio with a large, dry tunnel opening high in a vertical cliff face. Next to the south, the Rancho Nuevo is a huge karst plateau riddled with shafts which are largely unexplored. Lower down in the same limestone block lies the long and deep cave system of Purificación; clearly there is still major potential for new exploration in this area.

At the southern end of the state of Tamaulipas, the Sierra de Guatemala contains the Sótano de la Joya de Salas with fine shafts and a stream canyon reaching a depth of 376m. At the foot of the Sierra, two major risings, the Nacimientos of Río Sabinas and Río Mante have been dived to depths of around 100m. The Sierra del Abra is a huge karst straddling the frontiers of Tamaulipas with San Luís Potosi. At the northern end of the range, the Cueva da Diamante is a difficult cave to explore with a long, narrow canyon to a shaft system which descends abruptly to −621m. Further south, the Sótano del Arroyo is a more spacious system 7,200m long.

San Luís Potosí and Queretaro contain the heartland of the high karst with most of the deepest shafts and caves. The huge daylight shaft of Golondrinas is unquestionably the world's

SISTEMA PURIFICACIÓN

This remarkable cave system lies in the Sierra Madre Oriental, on the western edge of Tamaulipas. Now over 51km long, Purificación is 895m deep and offers the deepest through trip in the world, only 27m short of the system's total depth. Brinco is now just one of three entrances at the top of the system, which eventually leads down through active and fossil passages and out through the mazes of the Inferniello cave to the lower exit perched 20m up a cliff over the resurgence. The cave has enormous variety in its passages, with great lengths of horizontal tunnels, canals and lakes, as well as plenty of vertical features in fine, active streamways.

Brinco was first connected to Inferniello in 1978, and major new discoveries have extended the system almost every year since. Nearly all the exploration has been by American cavers.

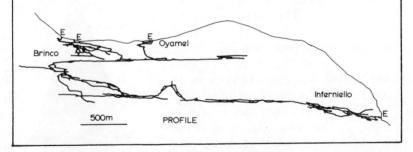

finest. In total contrast, the nearby Sotanito de Ahuacatlan has a narrow interior shaft of 288m in total darkness. Also in San Luís Potosi, the Sótano de Trinidad is a magnificent multiple-drop cave system reaching −834m in the deepest of its branches. Straddling the state boundary, the Xilitla plateau contains the Sótanos de Nogal (−529m) and de Tlamaya (−454m), both fine shaft systems. Also in Queretaro, the Sótano de Tilaco has a magnificent sequence of waterfall shafts reaching a depth of 649m. In contrast, the Sótano del Barro is a gigantic daylight pit offering a 410m rope descent, but it lacks the impressive overhangs of Golondrinas. Across in the state of Guanajuato, the Sótano de las Coyotas reaches −581m.

The southern end of the Sierra Madre Oriental reaches into the state of Puebla. In a karst of steep slopes clad in thick rainforest, the Sistema Cuetzalan is a dendritic system of large and long river caves. It includes the splendid river passages of Chichicasapan and Atischalla and reaches a total length of 22.4km and a depth of 587m. Similar river passages form the Atepolihuit de San Miguel (7,700m and 399m deep). Further west, near the village of Xochitlan, two parallel cave systems have fine long descending streamways laced with cascades and deep pools; the Sumidero Santa Elena is 7,900m long and 400m deep, while the Sumidero San Bernado is 3,900m long and 215m deep.

SÓTANO DE LAS GOLONDRINAS

This enormous and famous shaft lies high in the Sierra Madre Oriental, close to the remote village of Tamapatz. The surface opening is 50m across, and the shortest descent to its floor is a completely free hang of 333m. It was first descended by the Americans in 1967, and the much smaller continuation to −512m was found in 1969. A feature of the shaft is the population of swifts (*golondrinas*) which are always diving in and out of it. Golondrinas is no longer the world's deepest shaft, since Sótano del Barro was found in 1972, but the rope hanging right down the centre of the magnificent bell-shaped shaft still makes it the world's great vertical experience.

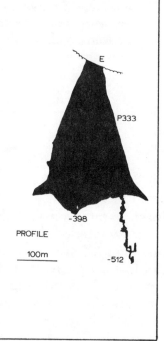

The Southern Mountains

Across the southern part of Mexico, structurally complex ranges of mountains traverse the states of Guerrero, Oaxaca and Chiapas before extending into Guatemala. They contain some major limestone areas with well-developed karst and extensive cave systems.

SISTEMA HUAUTLA

The Huautla Plateau, a 2,000m-high block of limestone in northern Oaxaca, contains one of the finest collections in the world of long, deep, wet and spectacular caves. Around the little village of San Agustín, huge depressions drain into the caves, of which the deepest is the Sistema Huautla, now 34km long and 1,252m deep. The main entrance to the system is down the staircase of wet shafts in San Agustín. At a depth −600m, these break into a long complex of large, dry fossil passages, with another series of streamways and shafts to a deep sump at −861m. Li Nita is the highest entrance, from where a long streamway, small and wet in parts, descends to the lower part of San Agustín and gives the total depth of 1,252m. The third way in is down La Grieta, another superb system of waterfalls and clean streamways.

Nita Nanta (12km long, −1,079m) and Sótano de Agua de Carrizo (3,750m, −836m) are two more spectacular shaft systems, both of which have been traced to boulder chokes close to the upstream parts of San Agustín. Despite many efforts, the connections have not yet been made, but they are only a matter of time. Yet more caves are intertwined and are integral parts of what could become a single huge system; they include Nita Nashi (−641m), Nita He (−594m) and Sótano del Río Iglesia (−531m). All the caves drain to the resurgence of Pena Colorada, where over 7,000m of cave have been explored beyond a 520m sump.

The Huautla Plateau was discovered by American cavers in 1965, and a depth of 612m was reached in 1968. Exploration resumed in 1977 and has continued with annual major discoveries. The potential is still tremendous.

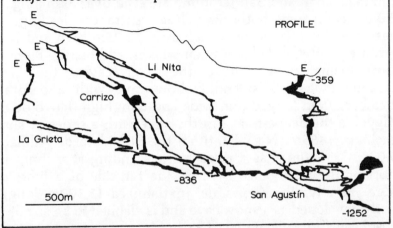

The Chiapas Highlands rate as one of the finest caving areas in the world, with numerous sinking rivers in a rugged, rain-soaked terrain. The caves are exceptionally fine, though none yet known is of very great length. The longest and deepest is the complex resurgence system of Veshtucoc, now with 4,900m of passage over a vertical range of 380m. Justifiably famous is the Sumidero Yochib, with a powerful river making its polished canyons, roaring cascades and deep plunge pools a serious proposition, though it is only 3,300m and 206m deep. The Cueva de la Chorreadero provides easier exploration in a superb through river cave 3,300m long.

A major karst region straddles the borders of Oaxaca, Puebla and Veracruz and includes the Huautla Plateau with its magnificent deep caves in northern Oaxaca. Across in southern Veracruz, the Sierra de Zongolica has many fine caves in a spectacular mountain karst. The Sótano de Ahuihuitzcapa has a 200m entrance shaft to a steep streamway which reaches a sump at −515m, and the Sótano Tomasa Kiahua (−374m) has a 330m entrance pitch. In contrast, the Sumidero de Atikpak is a gently descending river cave which reaches a depth of 320m. Over the border in southern Puebla, the Zoquitlan area has river caves with clean rifts and considerable vertical development, and the Canoajoapan area has many entrances waiting exploration in an area with a depth potential of 1,100m. Within this same region, recent exploration around Chilchotta has revealed a number of caves over 700m deep, including Guixani Guinjao (−940m).

The karst is extensive across northern Guerrero and contains some outstanding caves. The adjacent river caves of Chontalcoatlan and San Jeronimo are each over 5km long and 250m deep; they both have huge entrances leading into massive canyon passages which average 25m high and wide; they are gently graded, with many lakes, and feature flights of gours rising 60m. Above the river caves, the Gruta Cacahuamilpa is a splendid show cave with a decorated chamber 500m long, 120m wide and 60m high. Hoyo de San Miguel is the deepest cave in the state with a series of shafts reaching −455m. Near Pantitlan the Sumidero de Atliliakan (3,000m, −230m) has a river passage ending at a sump just short of the resurgence cave on the far side of a limestone ridge. Finally, the Gruta de Juxtlahuaca (5,100m long) is partly developed as a show cave and is claimed to be one of the most beautifully decorated caves in Mexico.

MONACO

The very small principality contains a beautiful show cave, the Grotte du Jardin Exotique. Very little other karst is recorded.

MONGOLIA

There are no known areas of karst within this country.

MOROCCO

Limestone is to be found to a greater extent in Morocco than in any other African country. It occurs in the mountains of Er Rif, the Middle Atlas, the Anti Atlas and the High Atlas and in many of the lower lying hills. The actual areas of karst are equally extensive but tend to be fragmented into smaller regions. Apart from the main mountains, cavernous rocks can be seen south of Tangier (of archaeological importance), in the low coastal hills between Safi and El-Jadida (Ghar Karker, −140m; Ghar el Goran, 1,600m long), inland from Essouiara (Ghar Isk N'Zouay, 2,300m long), west of Meknes (Ghar Bel Hordaifa, 1,200m long) and in the infrequently visited Monts des Beni (Oujda) and the desert region around Bou Denib.

In the Anti Atlas, limestone plateaux overlook the Sous valley but no karst has so far been recorded. The High Atlas, at its western end, consists of dissected limestone plateaux, and in the area just north of Agadir is to be found Morocco's longest cave, the the 11km Wit Tamdoun. Once past the igneous core of the High Atlas, karst has developed extensively amid spectacular scenery. The southern tip of the Beni Mellal karst just breaches this region, while the large Todra and Dades gorge district, the Plateau des Lacs and the Midelt area have been only sparsely explored.

Further north, the karst of the Middle Atlas continues as a series of tabular blocks. The Beni Mellal area is superb,

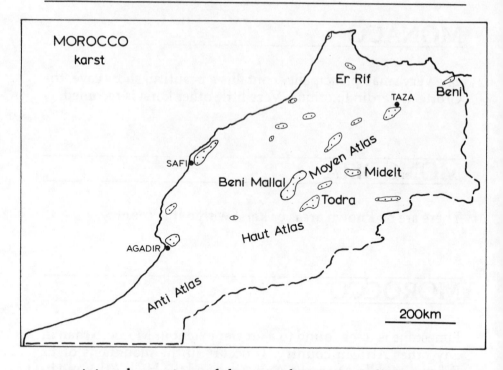

containing depressions, dolines, sinks, springs, lapies, shafts and numerous open caves (Ain Melghfi, −251m; Ifn n'Taouia, 3,600m long). Similar landscapes continue, until the end of the range is reached at the massifs above Taza. Because of its accessibility, this fine region has often been visited, its karst is well documented, and its caves − such as Kef Tikharbai (−310m) and the Grotte de Chara (8,000m long, with a very fine river passage) − are classics. The same area houses the Friouato and Chiker caves, each with around 2km of passage but still not connected.

The northernmost range, the highly folded Er Rif, possesses only a small area of karst relative to its size but does contain Morocco's deepest known system, Kef Toghobeit.

Early explorations of the caves were by both national and local clubs, but most recent exploration has been by French, British and Belgian expeditions.

KEF TOGHOBEIT

South of Chechaouen, and about 8km north-east of Bab Taza, lies Morocco's deepest system, at an altitude of 1,700m. It was discovered by cavers from Rabat in 1959, who explored to a depth of −377m the following year. Between 1969 and 1974 various foreign expeditions bypassed the then terminal syphon to reach the present depth of −677m and gain a total length of 3,600m. A 76m entrance shaft leads to a large passageway and many boulder-strewn chambers. After a complex, loose and rather muddy section around −300m, the character improves to give a sporting trip to the bottom. The water is thought to emerge 170m lower down, at an altitude of 850m.

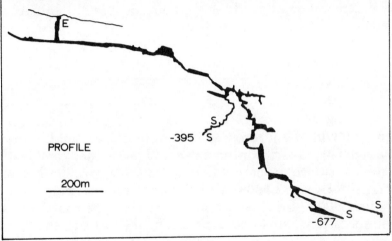

PROFILE

200m

-395

-677

MOZAMBIQUE

Few details are known of the hills in the western side of the country but on the large limestone plateau to the west of Beira, partly within the Garongoza Game Reserve, large dolines and disappearing streams have been recorded.

NAMIBIA

Large areas of limestones occur north of latitude 20 degrees and one region, the dolomite triangle formed by Tsumeb, Grootfontein and Otari, has been well documented. Most of the caves explored are partly flooded, and depths quoted often include considerable dives. The principal systems include Harasib (−152m including a 52m-deep dive from a spectacular chamber 125m by 35m by 60m) and Ghaub (1,070m long). One characteristic of the dolomite is the many cenote lakes, such as Guinas and Aigamus, which provide much sport for divers. Other less well-explored areas of limestone exist further south in the hills by Bethanic and Grootfontein. The cave explorations have been by both local and South African cavers.

NEPAL

Limestone has been recorded in various places, but it is only in the Kali Gandaki region, and around Halesi, that karst features have developed to any extent. In the former, the limestone occurs between 2,500m and 7,700m, but there is no karst above 4,500m. Small areas of karren are to be found together with occasional enclosed depression, springs and small caves. Close to Pokhara is the only long cave in Nepal – the Patale Chhango, with its spectacular river sink and 2,960m of passage. One other cave of note is Guptesway Gupha (Kusma), 190m long and formed in loosely cemented terrace sediments; nearby Alope Gupha is 500m long and 63m deep. The less-visited Halesi region is in the east of the country, but so far only small caves have been noted.

NETHERLANDS

No areas of karst are known to exist but limestones do occur in the southern province of Limburg.

NEW ZEALAND: Gardener's Gut Cave

OMAN: Kahf Hoti

PAPUA NEW GUINEA: Nare, New Britain

PAPUA NEW GUINEA: Selminum Tem

PERU: El Tragadero, Rio Chancay

SARAWAK:
Gunung Mulu National Park

SARAWAK:
Gua Air Jernih

SPAIN:
Cueva de Cellaron, Matienzo

TURKEY:
Tigris Tunnel

USA: Mammoth Cave, Kentucky

USA: Carlsbad Caverns, New Mexico

USSR: Zolushka

YUGOSLAVIA: Kacna Jama

YUGOSLAVIA: Divaca Jama

NEW CALEDONIA

This very mountainous Pacific island is richly endowed with a limestone karst of considerable but as yet unexploited potential. The two current major systems are the 3,900m Acho and Koumac, 3,700m. The Ile des Pins is also known to have a considerable number of caves.

NEW HEBRIDES

These Pacific islands have limited outcrops of limestone but no known caves.

NEW ZEALAND

The splendour and diversity of this country's scenery are fully complemented by an abundance of impressive caves and karst. The very active local cavers, in action for over two decades, already claim two of the deepest systems in the southern hemisphere as well as a number of others which exceed 1km in length – proof, if needed, that New Zealand stands among the world's major caving nations.

South Island

The great majority of the country's deep caves are found in South Island, mostly on the wild, rugged, mountain terrain of the west. There are many areas of limestone, but the principal karst region is that encompassing Mount Owen, Mount Arthur, Takaka Hill and Canaan. The altitude ranges from 1,778m down to 300m, and on the higher ground of Mounts Owen and Arthur the most striking feature is the barren, glaciated surface pitted by shafts. Unfortunately many of these are blocked by frost-shattered material. On Mount

NEW ZEALAND
limestone

Waitomo

Caanan

Paparoas

Mt. Owen

Te Anau

200km

Arthur, open shafts lead to both deep and extensive systems; five exceed 300m in depth, including Laghu (−307m), Blackbird (−316m), Gorgoroth (−346m) and HH (reaching 622m). The latter is a fine cave, originally investigated as a possible upper entrance to the Nettlebed Cave (+690m) but eventually proving to be a major cave in its own right, some 2km long and including pitches of 50m, 61m and 55m. The similar setting on Mount Owen has resulted in numerous steeply descending caves but none has so far exceeded the depth of the 291m Curtiss Ghyll. Two of the country's most exhilarating systems are to be found on the lower, but adjoining, Takaka Hill/Canaan area. Here, among a multitude of dolines, is the 176m entrance pitch of Harewood Hole, which then connects with the superb river passage of Starlight

NETTLEBED CAVE

The entrance to this very important system, in the Mount Arthur region of South Island, is situated several hundred metres up Eyle Creek, a minor tributary of the Pearse River. Discovery was made in 1969 and, after a kilometre of unexciting passage, exploration was terminated at a draughting constriction. In 1979 further investigation, following a little blasting, revealed an amazing complex of chambers, formations and some 6km of galleries leading steadily upwards for 260m. Subsequent and often difficult exploration has extended the system for more than 20km over a depth range of 687m (−29m, +658m). Much still remains to be done.

The main river within the cave, which is met only briefly, is known to emanate in part from Grange Slocker, a sink high on Mount Arthur. Between the Grange Slocker and Pearse Rising there is a vertical difference of 976m and a horizontal one of 6,500m. Exploration above the river's probable course has revealed a number of deep caves, including HH Cave now 622m deep; Wind Rift is already explored to −375m and may connect to Nettlebed to give a 1,000m-deep through trip. The Rising itself has been dived to a depth of 21m, but it is thought unlikely that any physical connection will be made with the main cave.

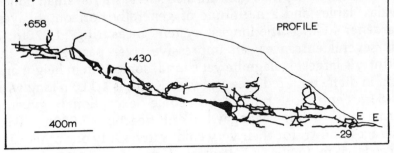

Cave to form a system of −357m and 1,310m in length; Greenlink Hole (−372m and 2,000m long) also presents a serious and sporting challenge to any caver.

The second important area is that of the western Paparoas hills and the Buller River region. This has been less well explored than the previous area due to hindrance by dense bush, but many extensive caves are known. These include the complex 5,010m-long Xanadu System and Metro Cave, with well over 8km of passageway. The Metro is of special interest due to an abundance of fossil galleries, in an area otherwise renowned for very wet and much younger caves. The arduous caves of Profanity and Damnation lie to the very north of this region near Inungahua.

The only other area of karst of note on South Island is that around Te Anau in the centre of the spectacular fiordland. On Mount Luxmore several kilometres of cave have been explored, but no individual one is of great length. Aurora Cave, nearby, is 213m deep but otherwise none is of impressive depth.

North Island

The physically very different North Island contains a comparable extent of limestone outcrop, but only Waitomo can claim to be a major karst region. However, its extent, its diversity and its profusion of caves fully compensate for the little to be found elsewhere. It extends from Kawhia Harbour in the north, inland to Te Kuiti and then down the River Mokau to its outlet into the sea, at Mokau. The land is low, rolling cattle country, never reaching 1,000m in altitude. The karst landscape is dominated by dolines 40-75m in diameter and 10-30m in depth. There are also sinks, shafts, small poljes, uvalas, lapies and a multitude of generally horizontal caves. Gardener's Gut is the longest system, measuring 11,270m; it has several entrances, an impressive river passage and the country's largest stalagmite yet found, some 10m in height and 12m in diameter. Fred's Cave, by Te Kuiti, is 4,110m long with a 53m entrance pitch and some exceptional gypsum formations, while Mangawhitikau reaches 5,900m, with passages known for their very cold water. Only one other cave exceeds 5km (Miller's Waterfall Cave, 5,150m) but seven others currently extend for over 2km. The region gets its name from the Waitomo Cave, a major tourist attraction and research centre; though its passages are short, they house an impressive population of glow-worms.

Lava caves are recorded on many of the volcanoes, such as Mounts Wellington, Albert, Eden, Smart and Rangetoko; none is of great length.

NICARAGUA

Numerous lava caves have been reported from the volcanic ranges of the west, but they have not been investigated, while in the north-west of the country an area of limestone also remains unexplored.

NIGER

Limestone is known on the plain of Talak but is of little speleological interest.

NIGERIA

There is little surface limestone within the country, and no major karst features are known. Quite extensive caves have developed in the sandstone of the Udi-Orlu uplands between Enugu and Onitsha. Several of these, reported to contain large lakes, running water and large chambers, are currently under consideration as a major tourist development.

NIUE ISLAND

A part of Polynesia, this 150 sq km coral atoll is riddled with caves and chasms. These occur both around the coast and inland, and all are renowned for their fine calcite formations. The longest known system is Anatola Cave (600m).

NORWAY

While cavernous limestones are relatively minor within the area of Norway's main landmass, they are common in Nordland, between latitudes 65 and 68 degrees. Within this region, outcrops generally occur as narrow bands and in many areas are of marble. At lower altitudes, the karst is often found in a barren but undulating countryside dotted with birch and

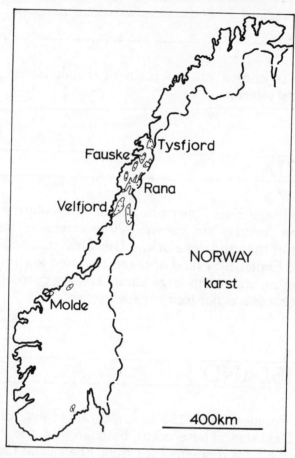

scrub. Dolines, depressions and dry valleys are the most common features, with limited, well-developed lapies. As the level rises, the trees and scrub are missing and the karst becomes much more exposed, but exploration is made more difficult by variable patches of snow. The caves commonly have fine active streamways cut in polished rock, but calcite formations are the exceptions rather than the rule. Many explorations have been by British expeditions, but the Norwegian cavers have become more active recently.

The main region is within the county of Rana, where over a thousand caves have now been recorded. There is a fine glaciokarst between the Svartisen ice caps, and within it Trudenhullet is a multi-level system 3,500m long. Nearby, Larshullet is 326m deep but rather uninspiring, while further east, towards the Swedish border, are more fine caves including Jordbrugrotten, 3,000m long. Further north are many stream caves around Saltdal, Gilkesdal and Beiarn, while close to Fauske the labyrinth of Okshola-Kristihola has 11km of mapped passage with a vertical range of 300m. Just south of Tysfjord, the Råggejavreraige system, 620m deep, offers a spectacular through trip, and near Sorfjorden the shafts of Salthulene contain 2,000m of ice-filled passages. South of Mosjoen, isolated limestone outcrops occur in the Velfjord region, extending into Bindal and Gråne counties. Serious exploration has started here only recently, but over a hundred caves are already known, including Sirijordgrotta (1,380m) and Øyfjellgrotta (1km long and 110m deep).

Isolated caves are also known known in thin limestone outcrops far to the north around Tromsö, on the islands Karlsöy and Reinöy, and to the south near Rindal, Molde (Trollkyrka, 500m long) and Skein.

OMAN

Though most of Oman is a barren desert, the mountains both north and south contain large areas of limestone. The highest mountains of the north, the Jabal Akhdar, are formed of limestone and dolomite with local relief of over 2,000m; few caves are known, and the splendid through cave of Kahf Hoti (5km long, −262m) appears to be an exceptional occurrence. Further east, the Selma Plateau has less relief but is a remarkable karst; it has a number of caves and shaft systems currently being explored, two of which reach depths around 400m. Also in the Selma Plateau is the Majlis Al Jinn, with three shaft entrances into a single chamber over 200m by

300m and mostly 100m high, which appears to be the second largest in the world.

There are many other limestone massifs, not yet explored for caves, within the mountains of the north. The Dhofar region in the south also contains extensive limestone hills. Tawi Atair is a massive sinkhole with a vertical descent of over 100m into a deep lake, with a short passage leading off to another chamber; it is only one of a number of shafts in the area not completely explored.

The cave explorations in Oman have been carried out mainly by American and British ex-patriates; there is virtually no access for casual foreign visitors.

PAKISTAN

Vast areas of limestone exist in this mountainous state, but only minimal cave investigations have been undertaken. In most of the Chitral district the hills are carbonate, which reconnaissance expeditions have to date found essentially non-karstic; further east, in Chilas, 6,600m up the Rakhiot Peak of Nanga Parbat, the world's highest known cave, is just 70m long. Other potential cave areas may be found in the border districts of Kurran and Waziristan, the central Sulaiman range of hills and the particularly massive limestone of the Kalat plateau.

PANAMA

There are no known areas of karst within this little-developed country except near Panama City, where several small caves, totalling 460m in length, are known as the Cuevas de Chilibri.

PAPUA NEW GUINEA

The limestone mountains of New Guinea, mostly hidden beneath a dense cover of forest and jungle, contain some of the most exciting cave-exploration potential in the world. Nowhere does the wild karst terrain provide easy access, and very high rainfalls make surface conditions frequently appalling. There are thousands of caves and some enormous underground rivers, but the full depth potentials, which could exceed 1,000m, have not yet been realized. Most exploration has been by visiting expeditions, mainly Australian, British and French, though there is a growing group of active local cavers.

Most of the mainland caves are in the highlands which form the backbone of the island. The geology is complex and,

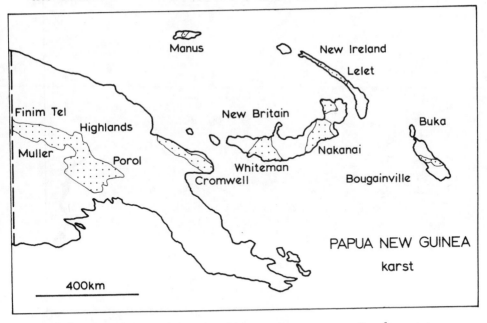

though karst is extensive, it is not continuous. In the eastern highland, the Porol escarpment has many caves, including Bibima, with a single passage 1,220m long giving an easy, steady descent to a sump at −494m. The Hindenberg Range, in the western highlands, has a karst plateau of huge dolines with many immature caves; Terbil Tem (−354m) is the deepest of numerous shaft systems. Within the Finim Tel plateau, Selminum Tem is 20,500m long with a huge, dry main tunnel and various lower streamways. To the south the plateau ends at the Hindenberg Wall, with a short fossil cave as a hole in the middle of it, but karst extends to the Star Mountains and elsewhere.

MAMO KANANDA

This huge cave system lies beneath the doline plateau of Mamo in the Muller Range. It has at least five entrances, including the huge Pugwa doline opening to a vast gallery. The passages now extend to 51,820m and reach a depth of 528m. They consist of a complex maze, on multiple levels, of fossil tunnels, flood routes and active streamways and contain three giant chambers. Both calcite and gypsum decorations are fine in parts. Mamo was explored by the Australian expeditions of 1978 and 1982.

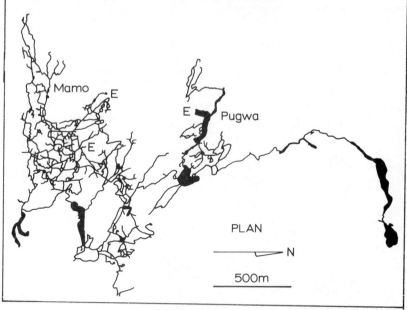

PLAN

N

500m

In the centre of the highlands, the Muller Range has some huge river sinks draining to the Nali rising, and also the Mamo doline plateau. The main river sink is the Atea, where four huge waterfalls crash into the one doline. Atea Kananda has a short but very impressive river passage to a sump at −92m, but tributary and side passages extend the cave to 34,500m long and 350m deep. Of many smaller potholes, the Uli Guria shaft system and the longer Heiowa Heia both reach −314m. Mamo has the large cave of the same name, and also Malemuli, whose series of shafts reach a sump at −420m.

Elsewhere on the mainland, the Tobio resurgence in the Kagua area has a mean flow of 100 cubic metres per second, but no open cave. The Cromwell Ranges have sinkholes in a doline karst, but only short caves are known, and a large doline karst near the south coast has low relief and little cave potential.

Of the offshore islands, New Britain is by far the most important, but karst also occurs on most of the others. New Ireland has extensive doline and cone karst. This includes the central Lelet Plateau, but the largest known caves are the single streamway of Lemerigamas (1,300m long, −203m) and the Dalum maze (1,200m long). Manus island has the river cave of Pumpulyun (1,250m), among more caves and some potential.

Bougainville Island has karst forming the central Keriaka Plateau; this contains the Koripobi cave with a 100m-high entrance into a single huge chamber 270m long and 150m high and wide. By the east coast, the Taroku Nantau Efflux is a resurgence cave 1,900m long in the Mentai karst. And on nearby Buka island, the Malasang cave is over a kilometre long. Many of the karsts of Papua New Guinea and its scattered islands have barely been examined for caves, and discoveries lie in wait.

New Britain

Limestone plateaux and active volcanoes combine to create a dramatic terrain on this forest-clad island. The most important karsts are in the Nakanai Mountains and the Whiteman Range, but there is also karst in the Raulei Range of the north.

Most of the Nakanai rises to little over 1,000m. It is a karst of giant dolines and shafts, with a number of deep gorges cut through it. The underground rivers are enormous, with huge catchments feeding to massive resurgences, unexplorable because of the power of the water. The cave potential is enormous, and most exploration to date has been in the more obvious large, open shafts. In the centre of the region, KAII is a fine pothole system (3,500m long, −459m), reaching a section of the huge underground river between the Kavakuna doline and the Matali rising. Further north, Minyé has an enormous shaft 350m deep, to a short but dramatic river cave; tributary passages make the system 5,500m long and 479m deep. Nearby are both the famous Nare and the beautifully decorated Gamvo river cave, recently explored for 6km to a depth of 478m. Also in the Nakanai, the plateau of Galowe is a major cave area rising to 2,000m; a 1985 French expedition discovered around 15km of caves. The major system is Muruk Hul, 4,500m long to a sump at −637m. South-west of Pomio the karst continues and contains Kururu, with a 200m shaft to 2,600m of passage.

NARE

Though neither long nor deep, Nare is one of the world's great sporting challenges. With a dramatic shaft dropping into a huge river passage, it is a caver's dream – though the size of the Nare river makes it more of a nightmare for its explorers. High in the Nakanai Mountains, the forest is punctured by the shaft which offers a free-hanging 230m pitch to the river. Upstream the passage is short, but downstream the size of the river makes exploration difficult and slow, with endless wall traverses and many dangerous crossings. The sump is reached after 2,000m, and the total system is 4,500m long and 400m deep. It was largely explored by the French in 1980 and finished by the British in 1984.

The upstream sump is just 400m from the Pavia cave (2,250m long, −260m). This contains another section of the Nare river and was found by the British cavers in 1985.

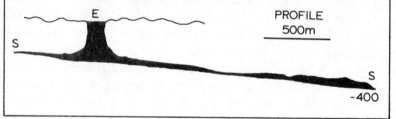

E

S

PROFILE
500m

S

-400

The Whiteman Range yielded 19km of passages to the French in 1985; it also contains giant dolines including Kukumbuone 280m deep with the Arrakis river passage at its foot. The Arrakis cave reached −468m and 11km in length, and the potential for future discoveries − in both the Nakanai and Whiteman − is clearly fabulous.

PARAGUAY

Limestone occurs only in small areas in the hills of the north-east, and some small caves were recorded by a recent French expedition.

PERU

This country, in common with most others in the Andes, is poorly endowed with areas of karst. Within its extensive boundaries, only two important cave regions are known. The lesser explored of these, but perhaps potentially the richer, is that in the north, within the fertile Cajamarca province. Here, among the tributaries of the rivers Chancay, Chotana, Chamaya and Maranon, many caves have been noted, and at least two extend for over a kilometre. The most notable are the five caves associated with Las Cuevas de Uchkupisjo, 2,400m long. In the centre of this area, by Cutero, the depth potential is greater, and the Tragadero de San Andrés, has been explored to −334m. The second major region is further south in the more mountainous province of Junin, and in particular around the town of Tarma. This has been the venue of several foreign expeditions to South America's deepest cave: Sima de Milpo is 407m deep and 2,140m long, mostly as a small, steep streamway. Many other discoveries have been made, but none presents a challenge to Milpo; most are of limited depth, but the main resurgence cave, Cueva de Huagapo, is 1,880m long. Tingo Maria, in Huanuca, is a further potentially promising area, but one which has so far failed to yield any caves of significance, while around the old Inca capital of Cuzco many small caves have been reported.

PHILIPPINES

All the larger islands and most of the smaller ones contain areas of little-explored karst. Only Luzon, Mindoro and Palawan have received any serious speleological study, principally because political and physical problems prevent exploration on the other islands. The potential is considerable. Most cave explorations to date have been by French, American and Australian expeditions.

On Luzon there are known to be more than thirty separate areas of karst but the three best known are those at Penablanca, Sagada and Libmanan. The former is situated on the north-western slopes of the Sierra Madre and contains the country's longest known system, Callao Cave, an incompletely explored river cave of over 9km. The Sagada karst is in the centre of the Cordillera Central and, while no very long caves have yet been found, it is renowned for its large dolines, up to 80m across, and its fine pinnacle karst. Numerous caves include the active system of the Latipan Cave (3,500m long, 160m deep) and Balangangan Cave (1,280m long), containing very large passages. The third and most recently explored area is on the south of the island, and, while no long systems have yet been discovered, the Calumpitan Cave has some 30m-by-30m passages. Mindoro has two large areas of virtually unexplored karst, one in the south and one in the north, to the west of Calapan. Here many promising sinks have been noted and several small caves explored. On Palawan, areas of dolines, karren and shafts are to be found in the centre and south. It also contains the incredible 7km-long river cave, St Paul's Cave, in which the first 3km have to be negotiated by boat direct from the sea; a through trip is possible.

Of the other islands, Samar is by far the most promising, with one large central karst area of over 900 sq km. Dolines, depressions, uvalas, lapies and caves have all been noted but not explored. Bohol and Cebu islands are both around fifty per cent carbonate, and Panay, Negros and Mindanao between them contain at least a further thirty karst regions. Coral islands, such as Borocay, commonly contain small caves with fine formations.

There are sixteen active volcanoes on the archipelago, together with numerous extinct ones, some very large. No lava caves are yet known, and they may be rare as most of the volcanoes are not basaltic.

PHOENIX ISLANDS

Low coral islands in the Pacific, with no known karst.

POLAND

The Tatras mountains, situated at the western end of the great Carpathian chain, provide Poland with a small but very important area rich in cave potential. The alpine peaks reach well over 2,000m in height, and, while some are composed of granite, the remainder are limestone. More than three hundred caves are known. These are generally to be found at altitudes around 1,500m, on the sides of the many deep valleys heading northwards towards Zakopane. Lapies and occasional bare dolines occur at the higher levels, while lower down, among scrub and woodland, dolines and depressions are common. Within this setting are many major cave systems including Jaskinia Wielka Snieźna, the complex and not fully explored Jaskinia Bańdzioch Kominioski (Kóscieliska valley, −570m, 8,700m long) and the country's longest cave, Jaskinia Miętusia (Mietusia valley, 9,015m long, 241m deep).

There are five further areas of karst, which, while of lesser importance, are none the less interesting. Of these, only the massif of Snieznika, in the beautiful Sudety mountains, contains any major cave systems. Jaskinia Niedźwiedzia is the longest, with 2,200m of passage and some fine formations. The Jura Krakowska, situated between the cities of Krakow and Czestowchowa, is a karst with a history claimed to include sub-tropical development. Spectacular limestone crags rise above rolling hills, separated by dry valleys. Depressions and dolines are common, while over five hundred small caves have been noted. The region includes the National Park close to Ojcow.

JASKINIA WIELKA SNIEŹNA

The deepest and most famous of Poland's caves is high up the Malej Laki valley in the Tatras mountains. The original Snieźna entrance, at an altitude of 1,701m, was first entered in 1969, when a depth of 377m was reached. A further expedition in 1961 followed the streamway to a syphon at −623m. While there are no large pitches, the system is wet and tight in places and contains loose rock in its upper reaches. In 1966 a higher entrance at a level of 1,872m was discovered; Jaskinia Nad Kotlinami connected with the older system below a series of meanders and shafts, increasing the depth to −755m. Subsequent diving in the syphon, during 1972 and 1974, reached more passages and a second syphon at −783m. Since then a third entrance, Jasny Awen, close to Kotlinami, has been discovered. Water from the cave emerges at Lodowe Zrodlo (altitude 947m) in the Kośeiliskiej valley nearly 5km away.

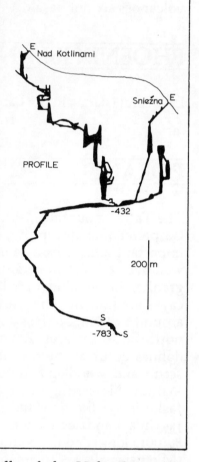

Around Kielce, in the gentle hills of the Holy Cross, are many more small caves, including the very beautiful but short Jaskinia Raj, one of the very few show caves in Poland. To the south of the Holy Cross is an extensive area of gypsum karst, including dolines, uvalas, poljes, sinks, shafts and small caves. Finally, between Lublin and the Soviet frontier there are chalk downlands supporting surface karst but containing no caves.

Polish interest in caves can be traced as far back as 1860, since when speleology, as both a sport and a science, has flourished. Hard experience in the Tatras, and a shortage of home karst, has stimulated Polish cavers into becoming worldwide expeditionaries with many notable explorations to their credit.

PORTUGAL

While over four hundred caves have been noted within the country, few are really extensive. The majority occur in the rolling hills between Coimbra and Setubal, with the Serra do Aire and Serra do Candieiros proving the finest karst (Gruta dos Moimbros Velhos, −180m, 3,000m long) The Almonda cave is 5,500m long. Dolines and sinkholes characterize these limestone hills. Coastal karst with excellent karren is extensive around Peniche and Sesimbra.

Many small caves are known in the band of limestone extending through the Algarve, and there is another small karst at the opposite end of the country, near Vimiosa.

PUERTO RICO

The island is predominantly of volcanic origin, but a belt of carbonate rock, up to 22km wide and 110km long, extends through the north-western region. High rainfall has created spectacular tropical karst which has been explored only partially. Cones, dry valleys, depressions, large dolines, sinkholes and shafts occur throughout, and over two hundred caves have been recorded. The largest are the segmented river passages of the impressive 9km-long Río Camuy system in the Lares limestone. The fine river cave of Río Encantado (9,600m long) is near Florida, and Cuevas Aquas Buenas, near Caguas, has 2,800m of passage. Most caves are very wet, but histoplasmosis is an ever-present hazard. Exploration has been carried out mainly by American cavers since the early 1960s. The nearby island of Mona is composed of limestone and is reported to be riddled with caves and shafts.

QATAR

This peninsular sheikdom is formed almost wholly of a low, crumbling limestone dome. A few springs remain as the only obvious karstic features.

RÉUNION

A wild volcanic island in the Indian Ocean containing many little-explored lava tubes. None has so far been found to exceed 350m in length, but the extinct and open crater of Commerson has been descended to a depth of 256m.

ROMANIA

With over 4,400 sq km of its surface area formed of limestone, Romania contains some of eastern Europe's most spectacular karst landscapes. The eastern Carpathians, which extend from just south of Brasov to the Ukrainian border, are composed mainly of broken massifs, usually covered in thick forests and crowned by high escarpments. Caves are to be found throughout, but the one outstanding district is that of the

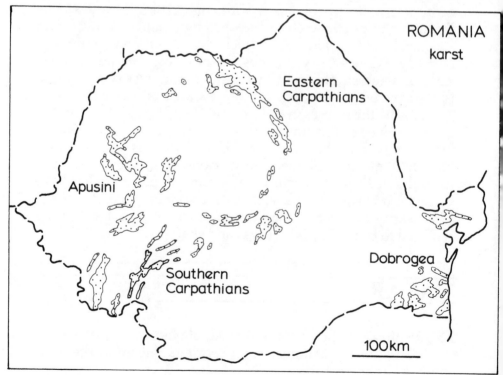

ROMANIA
karst

Eastern
Carpathians

Apusini

Southern
Carpathians

Dobrogea

100km

northerly Muntii Rodnei. This contains the deep Pestera de la Izvarul Tausoarela (−383m, 13km long) and the recently explored Pestera de la Jghiabul lui Zalion (−242m). The twisted tail of the Carpathian chain terminates in the much drier south-west as the southern Carpathians. Karst features there are superb, and poljes, dry valleys, ponors, dolines and karren are abundant. The caves are large, long (Pestera Topalnita, 20,500m) and deep (Avenul Dosul Lacsorului, −268m including a shaft of 155m), and some are very beautiful (Pestera Closani).

The true centre of caving is in the wet, wooded hills of the Muntii Apusini, where large rivers vanish and reappear with regularity within foreboding surroundings. Areas of particular note are Muntii Bihor and Muntii Padurea Craiolui, where many of the country's longest and deepest cave systems are situated (Pestera Vintului, 32km long).

The eastern karst of Dobrogea, while extensive, is of less importance. Low and fertile rolling hills have many small caves, but only one, Pestera de la Limanu, reaches any reasonable proportions (3,640m long).

Serious cave studies started in Romania around 1929, since when biospeleology has been particularly well researched.

RUSSIA

See Union of Soviet Socialist Republics.

RWANDA

No limestone karst exists but several extensive lava tubes have been explored by a Spanish expedition in the Virunga mountains. These include Ubuwume bwa Nyirabadogo (1,500m long, Bigowa) and Ubuwume bwa Muzanze (4,560m long and 210m deep, Ruhengeri).

SABAH

This small state, also known as North Borneo, has scattered limestone outcrops similar to those found elsewhere on the giant island. No serious cave exploration has been undertaken, except in the search for guano deposits and edible birds' nests. The largest recorded caves are at Gomantong, where Gua Simud Putih has fossil passages mostly over 30m wide connecting five horizontal entrances and a 140m-deep skylight. In most of the other limestone blocks, small caves are known, and the exploration potential is unassessed.

ST HELENA

Of no speleological interest.

SAMOA

While these islands are all basaltic volcanoes, only Upolu has so far revealed any lava caves. These are of some note because of their large passage dimensions, and the longest cave, Pe'Api'a on Mount Fito, reaches just 1km. The nearby Tokelau islands are low coral atolls with no known caves.

SARAWAK

Isolated outcrops of limestone are scattered through much of the larger Malaysian state on the island of Borneo. Though no individual karst is extensive, the high solution rates in the rain forest environment have produced karst landforms and caves on a monumental scale. Far the most important cave region is Mulu, but there are others.

Close to the northern coast, the Subis Hills have long been well known for the Niah Caves which are cut in them. Limestone cliffs rise 100m out of the alluvial swamps and fringe karst blocks which are riddled with caves. The Niah Great Cave has 3,200m of mapped passages, mostly of massive proportions, containing thick guano deposits and a vast number of edible swiflet nests. Nearby, Gua Limau has 8,500m of active and fossil passages. South of Kuching, isolated limestone hills in the Bau area contain extensive, mainly horizontal caves. Gua Jambusan (4km long) and Gua Tang Baan (5,800m) are complexes of large and small passages. Away from the coast, the limestone areas are, with the exception of Mulu, generally less well known, and large caves may await their first explorers.

Mulu

Within the Gunung Mulu National Park of northern Sarawak, a range of limestone hills dominated by Api and Benerat contains around 100 sq km of karst. Because of the different criteria involved, it is difficult to select the world's most spectacular area of caves, but Mulu must be a strong contender. The karst has 1,500m of relief, with towering cliffs, wild gorges, enormous dolines, rock pinnacles over 30m high, and huge cave entrances, all in dense rainforest. The caves are on a giant scale, and already over 150km of passages have been mapped, nearly all by British expeditions.

In Gunung Api, Gua Air Jernih is 51,600m long with a vertical range of 355m; nearly all its passages are over 25m high and wide and drain down into a massive river passage. In terms of sheer size, it is one of the great caves of the world. Other caves in the hill include Lubang Nasib Bagus with its giant chamber, and the beautifully decorated Gua Ajaib (4,770m). To the south, Gua Payau (2,160m) has perhaps the largest cave passage in the world, between 90 and 130m high and wide for its kilometre route straight through the mountain. North of Api, Gunung Benerat has the low-level maze of Lubang Sarang Laba-Laba (15,185m) and also the great high-level tunnels of Lubang Benarat (8,320m). At the northern end of the same mountain the River Terikan flows through four caves containing over 15km of passage. The exploration potential at Mulu is still dramatically high.

LUBANG NASIB BAGUS

This resurgence cave was discovered by the 1980 British expedition to the Mulu caves. It lies in the south-east corner of the Api massif. A fine tall river passage can be followed upstream, ultimately to a sump not far from the Hidden Valley sink. Total length of the cave is 2,900m, and it rises 423m. The remarkable feature of it is Sarawak Chamber, not far from its upstream end. This is by far the largest cave chamber in the world, around 700m long and 400m wide, and mostly around 100m high. The floor is a sloping mass of breakdown, and the roof is a gentle arch; both are devoid of calcite decorations. The river drains through the lower end, but its original route enters the upper end; it then cuts the chamber by slipping sideways down the bedding.

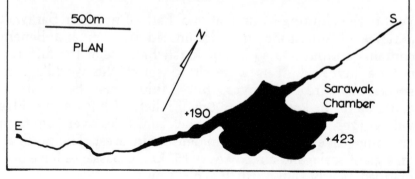

500m

PLAN

N

S

+190

E

Sarawak
Chamber

+423

SARDINIA

While there are several karst regions on the island, one area dominates all the others. This reaches up to 30km inland from the Golfo di Orosei, bounded in the north by Monte Albo and by the hills around Urzulei in the south. The dry terrain shows little obvious karst other than broken lapies, sinks, occasional dolines and some three hundred caves. These include the recently explored Grotta di Su Palu (Urzuli, nearly 12km long, +300m deep), Is Angurtidorgius (Nuoro, 8,800m long) and the island's deepest system, Voragine Filos d'Ortu, which reaches −319m with one pitch of 281m.

Of the other regions, the largest mass of limestone is between the Golfo dell'Asinara and Alghero. Karst features are rare, but caves do occur around the southern coast (Grotta di Nettuno, 860m long). The only other district of note is that just north of Iglesias, where the remarkable Grotta di San Giovanni is to be found. This system is 1,650m long, and for 850m a road traverses the large passage beside the underground river.

Most exploration has been accomplished by local cavers, but expeditions from mainland Italy, France and more recently Britain have also made significant discoveries.

SÃO TOMÉ AND PRINCIPE

Volcanic islands of no speleological interest.

SAUDI ARABIA

Most of the carbonate rocks in this desert kingdom are devoid of karst features. Exceptions do occur, the principal regions being found south of Riyadh between the 300km-long Jabal Tuwayq and the extensive escarpment of Al Biyadh. Exposed limestone depressions are not uncommon, and in the district of Al Kharj are the great dolines of Ain Samha and Ain Dhila, reputed to be around 100m deep and connected by underground drainage. The area also contains many shafts and small caves. More extensive areas of karst are to be found to the east of the Ad Dahna and in particular in the northern As Summan area; numerous shafts up to 30m in depth have been reported, and near Ma'agala one of many shafts leads into the Blowhole, with 600m of decorated passages.

SENEGAL

Of no speleological interest.

SEYCHELLES

Coral benches surrounding the igneous island cores support no significant karst.

SIERRA LEONE

Of no speleological interest.

SINGAPORE

Of no speleological interest.

SOCIETY ISLANDS

Rugged volcanic islands, the largest of which is Tahiti, with no significant limestone or caves.

SOLOMON ISLANDS

While this Pacific island group is predominantly volcanic, areas of well-developed limestone karst, much of them little explored, do exist on the islands of Malaita, Choiseul, Santa Isabel (Kolokafa, −80m) and Guadalcanal (Mbao Hol, 360m long).

SOMALIA

Limestone extends from both north-eastern Kenya and Ethiopia into the province of Upper Juba and again from Ethiopia into the north-west province. Karst has been recorded in both those countries but none has so far been noted in Somalia.

SOUTH AFRICA

While the total area of karst is minimal in relation to the country's size, its interest is enhanced by the variety of the rocks which contain caves. In the Cape Province, Table Mountain and the associated Cape Peninsula are formed of hard sandstones within which several extensive caves have developed (Ronans Well, 930m; Bats/Giants System, 460m). Very pure limestone supports well-developed karst in a small area east of Bredasdorp, and again north of Oudtshoorn in the Groot Swartberge. The former region contains many shafts and short caves which have been only partially explored due to the vicious bees which swarm around the entrances. The Swartberge, much larger and more amenable to exploration, includes South Africa's premier tourist cave, Cango, with its large chambers and 5,270m of passages. Nearby Eflux cave measures 1,760m. In the north centre, in a triangle roughly bounded by Kuruman, Vryburg and Campbell, is an area little investigated due to its dryness and inaccessibility. Dolines and shafts are known, often dropping to deep pools (Rushmangat, near Danielskull, has been dived to an unbottomed −80m). The Eye of Kuruman, on the edge of the Kalahari desert, is a large, unentered resurgence which has reputedly never been known to run dry.

Most of the karst in the Transvaal is found in massive dolomite which has been criss-crossed by hundreds of igneous dykes. Lapies and sinkholes are found on most outcrops, together with some 350 recorded caves. Cavernous areas to the east of a line drawn between Pietersburg and Pretoria are mostly on plateaux within the mountain country of the

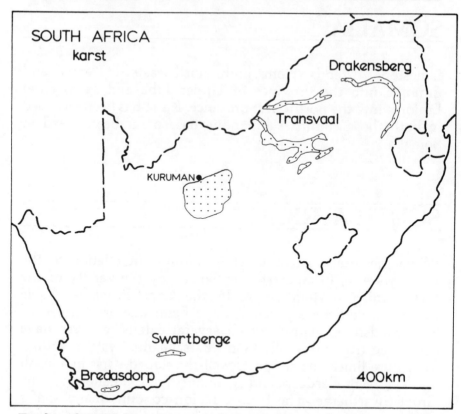

Drakensberg. Major caves include Wolkburg (−152m, 1,480m long) and the two tourist caves of Sudwala (1,800m) and Echo (1,720m). At the southern end of the region are caves and karst formed in quartzite (Berlin Caves, 1km). To the north and west of Johannesburg, the rolling hills provide a major caving area. Seven of the country's ten longest caves, many of them spectacularly decorated, lie in this one area. Apocalypse Pothole has a rectangular maze of passages over 11km long, and the decorated Westdreifontein Cave is 9,420m long. This landscape also contains many poljes and large dolines, and the Wonderfontein Valley is the site of gigantic and destructive sinkhole collapses created when the underlying gold-mines modified the dolomite drainage. North of Pretoria the long, thin arc of chert-rich dolomite outcrop supports minimal karst but does contain the very fine Thabazimbi Cave (4,480m long).

Recorded visits to Cango Caves date back to 1790, but most serious cave exploration has been since 1950, nearly all by local clubs.

SOUTH ORKNEY ISLES

Of no speleological interest.

SOUTH SANDWICH ISLES

Of no speleological interest.

SOUTH SHETLAND ISLES

Of no speleological interest.

SOUTH YEMEN

The high inland plateau, which stretches from the mountains of Yemen to Dhofar in Oman, consists principally of granite overlain by limestone. Much of this is again covered by loess or sand but caves are known to exist in the eastern Mahrah region. Surface karst features are also reported from the complex maze of limestone clefts and canyons which form the extensive central area known as Jol. Across the Gulf of Aden, the island of Socotra has dissected limestone plateaux around the central Haggier Mountains; caves probably exist there but none is yet recorded.

SPAIN

With more than twenty-seven per cent of its total area occupied by some form of karst, it is perhaps surprising that this country's true cave potential has only recently been fully recognized. The history of Spanish cave exploration dates back to the early 1870s, when investigations on an organized basis were commenced by the Centro Excursionesta de Cataluña and the Club Mantanyenc de Barcelona. Up to the

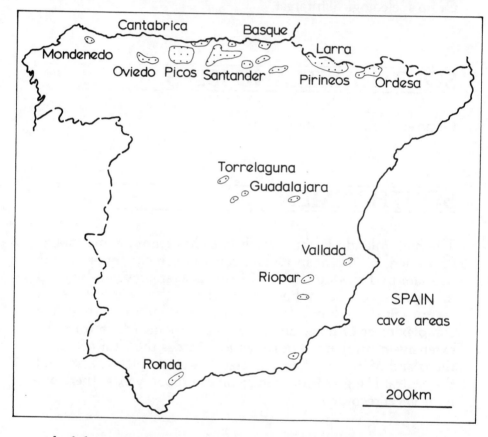

end of that century over thirteen hundred caves were recorded and at least partially explored. A period of relative inactivity then followed, enlivened by occasional, often amazing archaeological discoveries. After the Second World War, the local clubs again began their activities, but it was not until visits by various foreign clubs, since about 1955, that the enormous potential became apparent. A more serious and systematic approach to exploration was then initiated through the Comite Nacional de Espeleologia and its regional federations. Since that time, and particularly over the past decade, the results from the few areas explored are spectacular. Over fifteen thousand caves are now known, thirty-six of these exceeding 500m in depth and forty-five reaching 5km in length.

Two extensive regions currently dominate Spanish speleology, the Cordillera Cantabrica and the Pyreneean ranges. Almost all the known major caves are to be found here. There are also numerous other karsts, many of which are extremely interesting and of considerable importance in their

SIMA G.E.S.M.

1,670m up the Serranía de Ronda, in southern Spain, lies this very fine 1,098m-deep pothole. Currently it is the country's deepest system south of the Pyrenees and is the only one to exceed 200m in depth in the extensive karst of the Ronda limestone. The clean-washed cave descends via sixty pitches – mostly short, though one is 142m – over a distance of 2,700m to a syphon at −1,077m. This has been dived for 21m to give the present depth. The entrance in the depression of the Hoyos del Pilar was first noted in 1972 by the Group d'Explorations Soutérranées de Malága (hence GESM), and the sump was finally reached in 1978. It is considered a classic sporting trip with no great technical difficulties, though the many ascents make the return arduous.

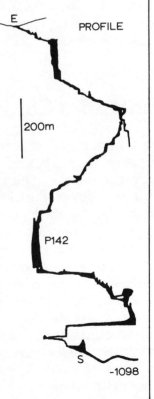

own right. In the latter category is the Serranía de Ronda, in the south. This consists of a deeply dissected, desolate tableland of well-developed karst. Caves are relatively few but include the 7,820m-long and 170m-deep through system of the Complejo Hundidero-Gato and the deep Sima GESM. Of equal standing is the Basque region, which forms a buffer between the Cantabrian and Pyreneean mountains. The wild hills contain many limestone karsts (Carranza, Durango, Ondarroa, Gorbea etc). The caves, many only partially explored, are in places impressive, though they never reach the dimensions of those in the neighbouring ranges. They include the Cueva de Mairuelegorreta (12,340m long) and Torca de Jornas (11,530m).

Other karst areas can be found in most of the eastern, central and southern provinces, with notable caves explored in Guadalajara (Cueva del Tornero, 11,000m), Torrelaguna (Cueva del Reguerillo, 8,270m, plus an extensive gypsum karst), Vallada (Tunel del Sumidors, 9,000m) and Riopar (Cueva de los Chorros, 14,600m).

Cantabrica Cordillera

The Cantabrians are a formidable mountain range extending for more than 300km from Lugo in the west to western Viscaya in the east, and from the coast up to 90km inland. On the lower slopes the setting is one of lush green meadows surrounded by thick, mixed forests. The higher levels, often spectacular with numerous jagged and snow-edged peaks, are wild and arid places, difficult of access but riddled with caves. These hide among great fields of lapies, at the bottom of dolines and in the sides of precipitous cliffs. The many short rivers plunge dramatically down to the sea and often cut impressive defiles.

The mountains of the west are the lowest and most complex and contain the least areas of limestone. Karst has developed in two areas of note. The smaller is the Valle de Mondenedo whose main interest is the five-level labyrinth of the Cueva del Rey Cíntolo (7,100m long). The main karst lies within a 30km radius of Oviedo. Its limestones are highly cavernous and culminate in the Sierra de Sobla (Cueva de Huertas, 14,500m long) and the Sierra del Gatto with its fine show cave of the Sistema Valporquero-Perlas-Covona. Many smaller caves exist just to the south-west of Oviedo and again around the Pena Candamo – some of prehistoric significance. Little limestone is then seen until the Río Sella, and the boundary with the most spectacular karstland of the Picos de Europa.

SISTEMA CUETO COVENTOSA

This 26,850m-long cave lies within the limestone of the Macizo de la Pena Lavalle, on the western side of the Val de Ason. Its depth (−815m) precludes it from a high place among Spain's deepest caves but the spectacular traverse, first accomplished in 1979, between its upper Sima del Cueto entrance and the lower Cueva de Coventosa exit, ranks it high among the world's most sporting systems. It comprises three levels, the uppermost of which is entered by a continuous series of pitches 580m in depth, including a single drop of 302m. This then joins the vast Galerie Juhue, which in turn is part of a 9km complex of great horizontal tunnels and enormous chambers – the Salle des Onze Heures covers 11,800 sq m. The upper series has one connection with the intermediate levels (between −590m and −740m), which are passages of a totally different character, narrow, twisting and often tortuous. The base level, that of Coventosa, has 10km of passage, dominated by the beautifully sculptured river gallery which is followed for 2km. The total traverse covers some 6km and descends 695m.

Sima del Cueto was first entered in 1958 by the French Spéléo Club de Dijon, but the main galleries were not reached until 1968. Cueva de Coventosa was investigated in 1954, but little exploration was carried out until a visit by the Spéléo Club de Paris resulted in 2,500m of passage in 1963. The resurgence, Cubera, is situated directly beneath Coventosa.

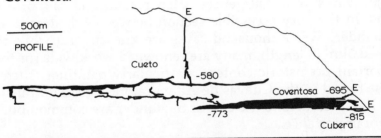

500m

PROFILE

Cueto

-580

Coventosa -695

-773

-815

Cubera

OJO GUARENA

This complex cave system is by far the longest in Spain, with over 85km of passages now mapped. A thick limestone forms the Cejo Cornejo escarpment, and a river sinks into it at the end of a deep blind valley. From there the cave extends as a multi-level complex of passages, with the main fossil trunk passage extending to the east. There are seven other entrances, including the large Palomera doline and the twin shafts of Dolencias which drop directly into a large chamber. Many of the passages are well decorated, and a feature of the main trunk is a series of long gour-dammed lakes; there are also ancient wall paintings near the Cubia entrance. New explorations have extended the cave annually for the last twenty-five years, and there is still considerable potential.

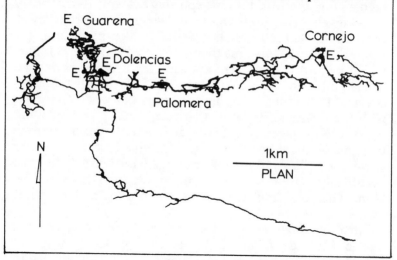

Between the Picos and the sea, and as far east as Castro Urdiales, is an almost continuous range of lower limestone hills. Karst features are numerous, with dolines, depressions, poljes, sinks and resurgences often reaching classic proportions. In the very accessible section between Ribadesella and Santander, over a thousand caves are known, and while few exceed 3km in length, many are renowned worldwide for their important prehistoric content, particularly paintings, such as those found in the Cuevas de Altamira, Pasiego, Tito Bustillo etc. Karst features to the east of Santander are common but of minor importance.

To the east of the Río Pas lies the more open and rolling relief of the Montañas of Santander. Karst again has a dominating influence on the landscape, creating a cave area almost as interesting as the Picos. While the often massive limestones are almost continuous, the area is usually divided up into three compartments – those of Ason, Ramales and Matienzo. The principal cave of the Ason region is the extensive Sistema Cueto Coventosa, though other major systems are being discovered and extended almost every year (Sistema del Hoya Grande, 15,300m, −471m). East of the Ason valley is the Ramales region, with the deep Cellagua cave system, and not far from Ramales itself, within the great

SISTEMA GARMA CIEGA CELLAGUA

This fine cave system is located east of the beautiful Ason gorge on the higher levels of the Martillano massif. There are two entrances, the Sima Garma Ciega (altitude 1,104m) and the Sumidero de Cellagua (altitude 949m). Both descend rapidly in series of short pitches to enter the massive main gallery – the former at −537m, the latter at −389. This passage extends for more than 3km and connects a series of boulder-strewn chambers before changing character at around −700m. From there to the bottom, the complexities of boulder chokes are met together with narrow canyons, squeezes and traverses, before the long sump is reached at −864m (originally wrongly surveyed to −970m).

Cellagua was first entered by French cavers in 1966. Then in 1969 the Garma Ciega was discovered, and a connection was made with Cellagua the following year. In 1975 the sump was reached at the same level as the resurgence in the Ason valley.

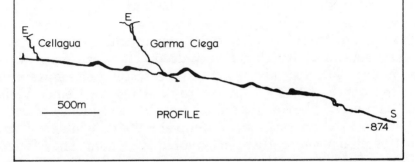

amphitheatre of the Arroyo Calera valley, is the 6,350m dry system of Cueva de Cullalvera. The hills to the north contain the Matienzo depression, an impressive closed valley covering more than 26 sq km. Within this and the surrounding hills, more than three hundred caves have been found, totalling over 60km of passages. The longest single system is the Cueva Uzueka (15km long). Spacious galleries and great chambers are a feature of the Matienzo caves, but there are also some fine streamways. On the southern flanks of the Cantabrians, a limestone scarp near Espinosa contains the Ojo Guarena, the longest cave in Spain.

Picos de Europa

This group of alpine-style peaks provide the highest (Torre de Cerredo, 2,648m), the most spectacular and the most cavernous part of the Cantabrian ranges. Massive limestones predominate, most of which have been folded and faulted. This has helped to create a wild and barren landscape, pitted with a multitude of impressive depressions, dolines, shafts, caves and similar karst features. Due to the rugged nature of the terrain and the fact that the area is a protected National Park, access is difficult, except around the perimeter and via four minor roads, with the majority of current and potential caving areas requiring long approach walks. Exploration has continued for almost three decades and, while four caves are now known to exceed 1,000m in depth, the true potential of the range is thought to have been barely touched.

Three separate massifs can be identified within the Picos, the western one being the Macizo del Cornion. This is bounded in the west by the Río Sella and in the east by the great Cares gorge. It has always been the most visited region and not unnaturally contains the greatest number of known caves, several of which are very deep. The Pozu del Xitu is 6km long with a steady descent in a cascading streamway, to reach a depth of 1,139m, and nearby the Sima de Cemba Vieja is 703m deep. The Pozu Cabeza Muxa has a sequence of deep shafts to a river passage and a sump at −906m, while the Pozo de Cuetalbo has recently been explored to −650m. The Macizo Central lies between the Cares gorge and the Rio Duje. It was thought to have less cave potential until 1982, when a French team explored the Torca Urriello; this has a steep sequence of shafts into a series of chambers which reach to −1,022m. Soon afterwards, the Sima del Trave was explored to a depth of 1,256m, and the Torca del Jou de Cerredo was followed to the top of a pitch at 744m.

Between the valleys of the Duje and Deva, the Macizo Oriental has a depth potential of over 1,500m. Tracks reaching up to old mines provide easier access across the wild karst

SIMA 56

Now the third deepest cave in Spain, 56 reaches a depth of 1,169m in a length of 4,900m. The cave is entered high on the magnificent Andara karst, in the eastern massif of the Picos de Europa. Most of it is a simple descending streamway, but its many narrow meanders and rifts and the succession of fifty-four pitches (mostly short, but two exceed 100m) make 56 an arduous pothole to explore. The terminal sump was reached in 1983 by a British team from Lancaster University. It is still 4km from, and only 180m above, the sump in the Agua resurgence cave. Though a through trip would be 1,500m deep, it is unlikely to be easily found.

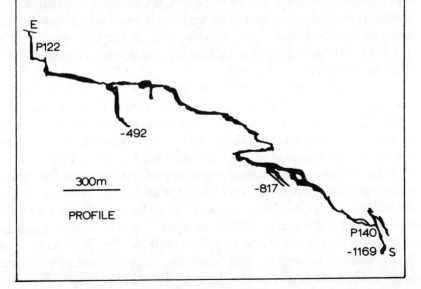

E
P122

−492

300m

−817

PROFILE

P140
−1169   S

terrain, and British expeditions have made some major discoveries. Sima 56 is the deepest; Dossers' Delight (−831m), Tere (−792m) and Flower Pot (−723m) are all steep shaft systems, and Sara (−648m) has a 280m shaft before descending more gently. The main resurgence is Cueva del Agua, where a complex, inclined fossil cave 10km long has been explored to a point 392m above its entrance.

Spanish Pyrenees

This great mountain range, which forms the Spanish border with France, is neither as spectacular as the nearby Cantabrians nor as beautiful as the French side of the Pyrenees. The much gentler slopes contrast sharply with the great rock faces of the French side, and a much-reduced rainfall has produced a less verdant landscape. Limestone and karst are, however, found extensively throughout the western and central regions, and the areas currently under investigation are yielding many major caves. In the west, the quite incredible Larra/Pierre Saint-Martin massif straddles the border to the south of the French village of Sainte-Engrace. Here, systematic research, still only partially completed, has uncovered two very deep systems, the Reseau de la Pierre Saint-Martin, mainly under the French side, and the Puerta de Illamina, entirely in Spain, as well as a host of others up to 800m in depth. Much more limestone exists to the west, south and east but, possibly due to less intensive investigations, no other major caves are known.

The central karst extends from the small town of Villanua in the west to the Río Cinca in the east. This is usually subdivided into three regions. The first, and most important, encompasses the great massif of Monte Perdido, the Parque Nacional de Ordesa and the hills around Escuain. Within this area there is perhaps the finest scenery of the Spanish Pyrenees as well as a multitude of karst features and caves; these include the difficult 1,149m deep Sistema Badalona and the recently explored S1-S2 system which reaches −846m and is 6,000m long. The Sierra de Tenenera lies west of the Río Ara; it is again very cavernous and contains the potentially much deeper Sistema Aranonera (currently 701m deep). Further west, across the great valley of the Tena, the limestone mountains around the village of Acumuer are dominated by a great field of lapies, in the midst of which lies the Cueva Buchaquera (−714m). But this and the rest of the central karst can be considered to have been only minimally investigated, and many areas exist where there is the possibility of caves exceeding 1,000m.

PUERTA DE ILLAMINA

This recently discovered cave, also known as BU56, is entered at an altitude of 1,980m in the rugged Larra region to the south of the Pierre Saint-Martin massif. It consists of a rapid succession of pitches to −433m and then large chambers and passages leading to a fourth sump at −1,338m. There are very fine formations throughout and a good-sized river in the main gallery. The water rises at the Emergence d'Illamina not far from the Kakouette Gorge in France − 9km distant and 200m lower. There are also possibilities of higher entrances being found. The very rapid exploration was made by largely French teams.

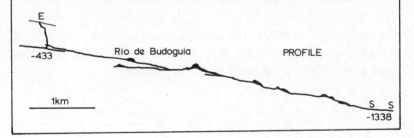

SRI LANKA

Karst is developed in two contrasting regions. In the north, on the barren Jaffna peninsula, Miocene limestone, barely exceeding 50m in altitude, is riddled with dolines, small shafts, lapies and many short but well-decorated caves. Further south, narrow bands of limestone and marble stretch from coastal

Trincomalee through the central highlands to Matara. Dense rain forest makes exploration difficult, but in the Kandy and Adam's Peak district there are many known caves (Istripura Cave, 600m long). Close to Ratnapura is a 200m-long cave formed in conglomerate, the Waupane River Cave.

SUDAN

Of no speleological interest.

SURINAM

There are no known areas of karst.

SVALBARD

Limestone is quite extensive on Spitsbergen, the main island of Norway's Svalbard group, and several areas of small dolines, karren and short caves have been recorded.

SWAZILAND

Of no speleological interest.

SWEDEN

Some fifteen hundred caves are known and, while the majority of these have formed in limestone, many are in other rocks such as granite and gneiss. Some of the latter are quite remarkable and include the 2,600m-long complex of covered granite fissures comprising Bodagrottorna close to Iggesund, the 180m-long granite block labyrinth of Rovargrottorna, and the permanent ice cave of Isgrotten by Frosen. Of the limestone karsts, various minor but interesting outcrops occur throughout the south and centre, but all the major regions are either on the island of Gotland or by the north-western border. Gotland contains some seventy-caves up to 3,100m in length (Lummelundagrottan), many of which are well decorated. To the north, Stara Blaasioen is notable for its caves in marble, and Strimasund for its many long caves (Labyrintgrottan, 2,100m; Sötsbackgrottan, 1,750m). The least accessible region and also the most promising is that on both sides of Lake Tornetrask. Already over a hundred caves have been recorded in the beds of marble, including the 136m-deep Kappasjokk-grottan and the 1,300m-long Lullehatjårrogrottan.

SWITZERLAND

With over four thousand caves recorded, forty-eight exceeding 1km in length, seventy-four exceeding 150m in depth, and five over 500m deep, Switzerland can be considered a major caving country. Limestones dominate in belts across the country, and particularly fine areas of karst occur in the Jura mountains of the north-west and in the long, rugged ranges of the Oberland, along the northern fringe of the main Alpine chain. Between these two regions, plateaux of limestone are limited, and few areas contain caves or karst of any note. The highest mountains of the Alps are mostly of non-karstic metamorphic rocks, but an area around Cavergno, in the canton of the Ticino, contains a number of caves in limestone (Acqua del Pavone, −158m, 2,900m long). Near Sion, there is a small gypsum karst containing the Grotte de Vaas (1,340m) and also the Saint Léonard cave with its large underground lake.

The rolling hills of the Jura are covered by forests and pasture. Karst is developed with depressions, dry valleys, dolines and many very fine river caves. With their easy access, these caves were the first explored and are the best known in Switzerland. Notable systems include the fine, long stream-

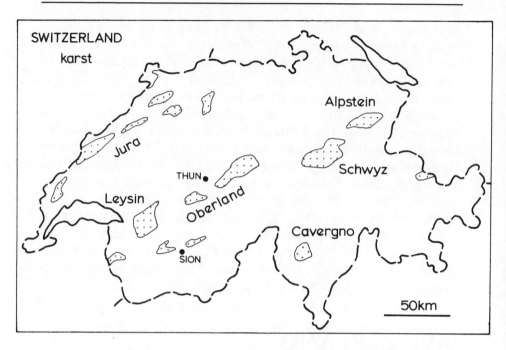

SWITZERLAND
karst

Alpstein

Jura

THUN

Schwyz

Leysin

Oberland

Cavergno

SION

50km

ways of the Grotte de Milandre (Boncourt, 10,520m long, +135m) and the more recently explored Nidlenloch (Oberdorf, 7,330m long, −418m).

The much higher Oberland, rising to 4,274m, has yielded, and continues to yield, the really major caves. The surface karst is rugged and often spectacular, with fine fields of karren, dolines, depressions, sinks and dry valleys extending over large areas. Important caves are developed in the Alpstein area, but even more so in the massifs between Glarus and Schwyz; these contain the famous Hölloch, and other caves such as Windloch (7,100m long). Just north of Interlaken and east of Thun, the massive limestone escarpments contain a number of spectacular caves. The largest is the Siebenhengste system, with eight entrances into a complex of ancient tunnels and a series of low-level streamways (now 80km long, 912m deep). The limestone around Leysin supports more fine alpine karst and also has the Chevrier cave (−622m).

Most exploration has been by local cavers, and there is still great potential, particularly in the central Oberland.

HÖLLOCH

This vast Swiss cave system has its lower entrance at an altitude of 734m, not far from the village of Muotathal, in the Alps of Schwyz. Passages lead both downwards, to a syphon at −116m, and upwards, into the main system. The 133km of passages are dominated by large galleries; they form two very complex inclined maze series, with the lower main series rising to only +296m. The fragmented upper series includes passages which follow the dip up to levels of +760m, giving a total range of 876m. Exploration of the system is made difficult by the labyrinth of passageways and the severe flooding which occurs at the lower levels; visits are normally made only during the winter months. The karst plateau above the cave is currently being investigated in an attempt to find further entrances. Promising discoveries so far include the Schwyzerschacht (−442m, 13km long) and the Discoschacht (−261m). The water resurges at an altitude of 638m, not far away, at Schleichenderbrunnen.

The cave was first entered in 1875, but major discoveries started around 1949, when a special exploration group was formed. Work in the cave continues today, and the potential is considerable.

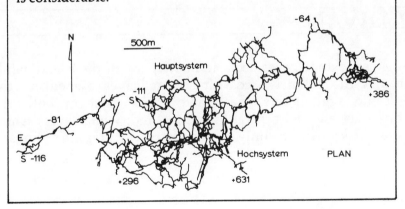

SYRIA

With only one known exception, all Syria's karst lies in the mountains of the west. Large springs, depressions and small caves have all been reported in the rolling Jebel el Ansariye, east of Latakia, while in the Jebel el 'Ala, close to Hama, an arid limestone plateau is pitted with depressions and has been cut by the deep gorge of the Nahr el Asi. The area with greatest potential, but least explored due to political considerations, is the Jebel esh Sharqui, which forms the boundary with Lebanon. Dolines, springs, depressions and cave entrances have been noted as far south as the Jebel esh Sheikh.

Away from the west, the only notable karst feature is that of Ras-el-Ain close to the Turkish border. This is a group of thirteen springs, all within a kilometre radius, which discharge an average of 38.6 cubic metres of water per second from an 8,100 sq km collection region and form the head of the Khabour river. No caves are known.

TAIWAN

Small sandstone caves occur in many parts of the island, while around the extensive coastline more caves are cut in the volcanic agglomerate. Carbonate rocks exist in both the north-east and the central highlands and, while several short caves are known, the only obvious karst features are springs.

TANZANIA

The only recorded regions of limestone karst are in a 150 sq km area west of Tanga, where there are very fine karren, dolines, small shafts and caves, and in the hills around Kibata (Grotte de Nduli, 3,500m long). Lava caves exist on Mount Kilimanjaro and could possibly occur on other northern volcanoes.

THAILAND

The karst of this beautiful country is both extensive and varied. It is developed in numerous small areas dotted throughout the central and western hills. Current research suggests that no caves of great depth or extent will be found, but potential for small to medium-sized systems is considerable. Its variety includes almost all forms of tropical karst, with caves of palaeontological religious and economic importance.

In peninsular Thailand the karst is at its most extensive and very similar to that of peninsular Malaya. Many small limestone hills rise steeply from the thick rainforest and exhibit well-developed features such as karren, dolines, dry valleys, small shafts and caves. The western coastal karst is

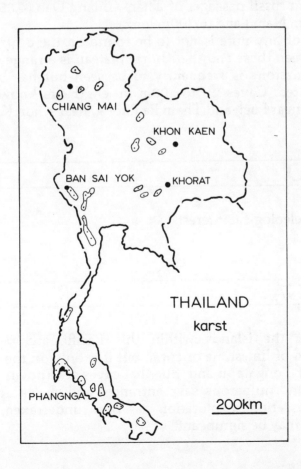

particularly spectacular in the Phangnga and Phuket region, where tall, eroded towers rise dramatically from the water. The longest-known caves of the peninsula are found here, such as Tham Ruesie and Tham Pong Chang, both 500m long. Many caves are very beautiful (Tham Lue Sri Sawan), and others are homes for the edible nests of swiftlets (Tham Ko Phi, Krabi).

The mountains of the west and north are generally composed of high granite ridges, some flanked by small outcrops of limestone karst. In the south, between the hills west of Phethaburi and those near Ban Sai Yok, karst is extensive and quite different from that on the peninsula. Several areas of cockpit karst occur, together with dolines and many small caves (Tham Kaeng Lawa, Kanchanaburi, 700m). Little-explored limestone karst is again found in the difficult mountain country within a 150km radius of Chiang Mai. The country's longest caves are found here, including the large active and fossil passages of Tham Chiang Dao (4,850m long) and Tham Nam Lang (6,700m long).

Karst of any note is not to be found on the central plains, but between these and the Khorat plateau is a range of rolling hills. Limestone is frequently to be seen but has been little investigated. Caves up to 500m in length are known (Tham Erawan, near Loei, and Tham Pa Puong, near Khon Kaen).

TOGO

Of no speleological interest.

TONGA

Many of the islands within this Pacific archipelago are composed of limestone or coral, but it is only on the principal islands of Tongatapu and 'Eua that caves are known. Here are to be found numerous cave entrances and shafts up to 80m deep. No serious exploration has been undertaken, but the potential may be significant.

TRINIDAD AND TOBAGO

On the island of Trinidad extensive areas of limestone karst are to be found in both the northern and central highlands. The region is rugged and densely forested, making exploration very difficult, but dolines, sinks, dry valleys, resurgences and caves are known. Those caves so far investigated are relatively small but frequently house the unusual Guacharo oilbirds. No karst is recorded in Tobago.

TRISTAN DA CUNHA

Of no speleological interest.

TUAMOTU

Pacific Ocean islands of low coral with no known karst.

TUNISIA

In the north of the country, the majority of rocks are limestone in the High Tell and the high and low steppes. Caves occur throughout, but the main areas are in the steppes. Ghar Djebel Serdj (267m deep, 1,700m long) and the Ain et Tseb resurgence (3km long) are currently the deepest and longest caves in Tunisia, both being in the Djebel Serdj. Ghar Kriz (1130m long, −150m) is in the Téboursouk mountains, and Ghar Zaghouan (−77m) is in the Djebel of the same name. There are many other smaller caves, springs and depressions.

TURKEY

Over 150,000 sq km (nineteen per cent) of Turkey is composed of carbonate rocks, most of which have developed karst to some degree. With certain notable exceptions, most of this vast area has been only superficially investigated. Several thousand systems are known, and while the scope for caves deeper than about 400m is minimal, the potential for lengthy caves is considerable.

Interesting areas occur throughout both Asian and European Turkey and culminate in the magnificent karst of the Toros Daglari. Minor karst regions, not necessarily fully explored, occur in the rounded hills to the west of Istanbul (depressions, dry valleys and caves up to 700m in length), north-east of Izmir (small caves, dolines and karren) and in the hills south of Mugla (poljes, depressions, springs and small caves). More extensive karst is found in the Bey Daglari, west of Antalya, but while small poljes, dolines and springs are quite common, there are no caves of any great size (though the sea cave of Kapitas Deniz has a main chamber measuring 75m by 60m by 20m high). Around Bursa are many fragmented areas of limestone, and in one, above the village of Ayva Köyü, is the 5,500m-long cave of Ayvaini. The major region outside the Toros Daglari is inland from Zonguldak, on the Black Sea. Recent explorations in these rolling hills have yielded the 6,250m-long river cave of Kizilelma, the 2,970m Gökgül and the 2,700m long cave of Atei Ini. Much more research still remains to be done in this area.

Further east, the limestone has been little visited but karst is known to the south of Trabzon, east of Agri and Lake Van, and again by Hakkari. The region around Lice contains the impressive through river cave of the Tigris Tunnel (860m long), together with poljes, sinks and caves in the nearby hills. Caves have also been reported in the Goksun area, and one expedition found small caves in the vicinity of Antakya.

An extensive area of gypsum karst is situated to the east of Sivas and contains small caves (Kocabey, 300m long), shafts, dolines, depressions and springs.

Serious caving in Turkey dates back to about 1960, and the national society was formed soon after, but much of the major exploration has been by foreign expeditions.

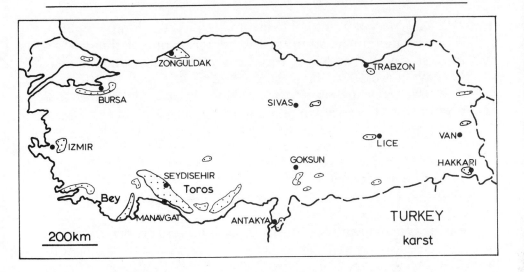

Toros Daglari

Almost ninety per cent of the known caves of Turkey occur in this southern mountain range, and most of these are to be found at its western end. East of the Gök Cay river valley, the limestone, often very massive, extends almost as far as the Ala Daglari. The karst features, however, tend to be poorly developed, and the potential for extensive caves is not considered great. The two notable cave regions of the east are, first, between the Cilician Gates and the Cakit river, and then to the north of Silifke (Cennet obrugu is −116m deep, and the nearby Cehenneum magarasi is a 90m shaft, 50m in diameter).

The limestone to the west of the Gök Cay forms a spectacular karst in the area between Ak Dag and Dedegöl Dag, around Seydisehir. The most remarkable features here are the great poljes and their long underground drainage routes (Beysehir, Sugla, Kembos, Eynif and Akseki). The whole area is rich in enclosed depressions, dolines (Kayaagil cuckuru above the Akseki polje measures 500m by 300m by 160m deep), large karren and some massive springs (the Dumanli rising has a base flow of over 25 cubic metres per second but now lies beneath the Oymapinar reservoir, upstream of Manavgat). The many fine caves and potholes include the country's deepest, the sporting, clean streamways and shafts of Düdencik (Cevizli, −330m), the 192m-deep shaft of Dünekdibi obrugu (Cimi) and the aqueous resurgence cave of Pinargozu in the Dedegöl Dag (5,270m long, +248m). Also situated in the south of this region is one of the world's longest caves formed in conglomerate: Tikiler düdeni near Manavgat is 6,600m long, −66m +93m.

Karst terrain extends westwards as far as Burdur and also occurs again to the north of Konya.

TUVALU

Pacific Ocean islands of low coral with no known karst.

UGANDA

Carbonate rocks occur in several small areas but no karst features are on record. Many small lava caves are known on Mount Elgon and could also exist on the other volcanoes of Karamoja province.

UNION OF SOVIET SOCIALIST REPUBLICS

Within the world's largest country, limestones cover more than 4 million sq km − about eighteen per cent of Russia's area. While only a small proportion of this has been thoroughly investigated, it is known that the greater proportion of these rocks support karst to some degree. This can be found in an amazing diversity of situations and conditions from sea-level up to over 3,500m, and from the frozen tundra of the north to the blistering deserts of the south.

Interest in the exploration of caves has increased dramatically during the past two decades, with the number of recorded systems growing from barely five hundred to a total

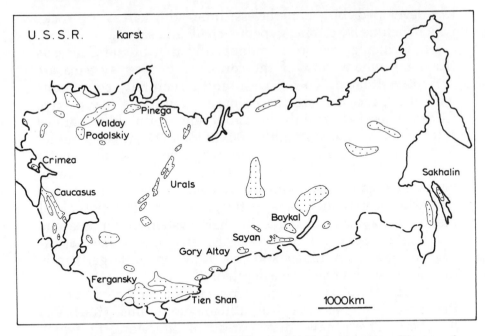

now exceeding five thousand. Thirteen of these extend beyond 500m in depth, and twenty-three reach more than 5km in length. Because of the vast areas still unexplored, the opportunities for new discoveries are considered greater than in perhaps any other country, and future exploration of a cave system, 2,000m deep is a distinct possibility. Already many of the caving districts of the Soviet Union – Podolskiy, Crimea, Caucasus and the Tien Shan – can be classed as major caving regions in their own right, while others must be considered important by any standard.

Work on karst commenced in the eighteenth century, when studies were instituted to improve agriculture and combat soil erosion. Now there is a very vigorous National Association of Soviet Speleologists, which covers all aspects of both sporting and scientific speleology to a very high standard.

European Russia

Due to its relatively high population and ease of access, the karst of this part of the Soviet Union has been investigated to a much greater extent than that of the Asian sector. Several major cave regions are known, together with many less-developed but nonetheless interesting areas in rocks which include limestone, gypsum, chalk and salt.

The principal region is undoubtedly that of the Caucasus (Kavkaz), where most of the country's deepest systems are currently situated. To the west, and structurally continuous with the Caucasus, are the Crimean mountains (Krymskiye Gory). These are mainly limestone and form three parallel ridges within an area roughly 100km long, 32km wide and up to 1,500m in height. The main range contains spectacular karst, with the deepest caves occurring on the high, lapies-covered plateaux known as Yaylas. Here are to be found the 500m-deep Soldatskaya system, the 287m-deep pothole of Nahimovskaya, the 270m-deep shaft system of Druzba and many others. The only long cave, Krasnaya, is situated at the edge of the main range and has 13,300m of large, gently inclined passages reaching a depth of 320m.

The structural zone of the Crimea-Caucasus continues to the north-west as the Carpathian mountains (Karpasky Khrebet). These have many areas of limestone in which surface karst is common, but while caves are to be found, they are generally small. On the edge of the limestone, in the upper Tissa river basin, is one of Russia's finest districts of salt karst where many sinkholes, dolines, shafts and even small caves have developed.

In the area known as Podolskiy, in the western Ukraine, lies the world's most important area of gypsum karst. Rarely reaching 50m in thickness, sinks, depressions, springs and caves have been developed to an extent unknown elsewhere. The many long cave systems include Optimisticheskaya, Ozernaya, and the recently explored Zolushka, with more than 80km of passages, some of which are notably large.

A very large but much fragmented area of limestone and gypsum karst is situated between Moscow and Leningrad, commencing with the Valdayskaya Vozvyshennost' in the south and extending as far as the Severnvy Dvina and Pinega rivers in the north. The longest caves again occur in gypsum and include Kulogorskaya (7,190m), Konstitutsionnaya (5,880m) and Olimpiyskaya (5,500m). In addition there are a further twenty-two caves each with a development of over a kilometre of passages.

Separating the great Russian plain from the even greater Siberian plain are the ancient, eroded and rounded Ural mountains (Uralskiy Khrebet). The range can easily be crossed as many large valleys are cut through the 1,800km length. In the central regions the country is heavily populated, and numerous mines scar the hillsides. To the north the wild and barren tundra remains relatively untouched, while to the south

OPTIMISTICHESKAYA AND OZERNAYA

The second and fifth longest mapped caves in the world lie next to each other in the lowland gypsum karst of the Podolskiy region, north of the Black Sea. Ozernaya was discovered in 1940 and now extends to a length of 105.3km; Optimisticheskaya was only found in 1966 but already has 151.3km of known passages.

The caves are formed in thin beds of horizontal gypsum. Each cave has only one entrance, and both systems are two-dimensional mazes of incredible complexity. Most passages are only a few metres high and wide, though they locally coalesce into small chambers. Though the longer of the two, Optimisticheskaya is contained within a smaller area than Ozernaya, and its survey is impossible to reproduce on a small scale. The two caves are only 950m apart, but as they are separated by the valley of the Korospeska stream, it is unlikely that they will ever be connected.

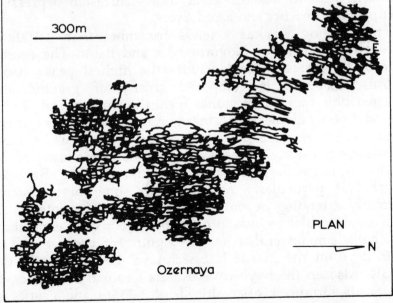

300m

PLAN

N

Ozernaya

a much lusher vegetation clothes the slopes. Both limestone and gypsum karsts occur and can be found as far north as the Par-Khoy Khrebet and as far south as the city of Bredy. The major cave areas are central, with that near Verkhneuralsk being perhaps the finest; it contains the 70m entrance shaft of the Sumgan-Kutak which leads into 9,860m of passages. Nearby are Zlebodarovskaya (2,850m) and several other systems exceeding 2km. The area has many caves rich in palaeolithic material. The second major region, though not so large, occurs in the gypsum around the cities of Perm and Kungur. The principal cave system is that of Kungurskaya, a show cave 5,600m long containing many spectacular ice formations.

Many other minor karst regions exist in the European section but none contains any caves of note.

Caucasus (Kavkazskiy Khrebet)

This spectacular complex of high mountains crosses the large isthmus separating the Black Sea from the southern Caspian Sea. It is divided into two ranges, the Bolshoy (Greater) Kavkas of the north and Malyy (Lesser) Kavkas of the south. Between the two lies the great trans-Caucasian depression occupied by the Kura and Rioni rivers.

The Bolshoy Kavkaz extends for more than 1,100km, between the towns of Novorossiysk and Baku. The central core of the range, which includes the highest peaks (Gora El'brus, 5,633m), is composed chiefly of granitic and metamorphic rocks. Limestones form the western end of this core and occur along the northern and southern flanks of the chain. Karst is extensive throughout and is at its most impressive in the great ranges of Abkhasia where many of the Soviet Union's deepest caves are known. The Bzybsky Khrebet is particularly notable, with some nine systems currently extending beyond 400m in depth. The potential is still considerable, as was proved by one recent expedition which made no fewer than seventy significant discoveries. The main cave on the massif is Snezhnaya, the deepest in the USSR. Also on the Bzybsky is Russia's fourth deepest cave, Napra. Its entrance is at an altitude of 2,335m, and a series of

SNEZHNAYA

Now the second deepest known cave in the world, the Snezhnaya system is entered high on the Bzybsky massif of the Caucasus mountains. In the Georgian language it is Schachta Towiani, and the name means Snow Cave; it was first explored in 1971 by Moscow University cavers, down through the vast snow banks of its doline and entrance shafts. The explorers were first stopped by massive boulder chokes around the junction with the main river gallery. Then in 1981 one passage was explored upstream, and the other was followed down a gentle gradient to a depth of 1,335m.

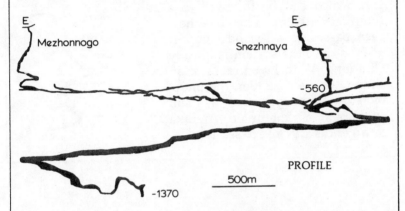

PROFILE

The Mezhonnogo pothole was explored down its own sequence of shafts, until it also met a large river passage which was followed downstream to a boulder choke. In 1983 this was passed into the upstream gallery of Snezhnaya, bringing the total depth to 1,370m. The cave is now over 19km long, and there is still potential for further discoveries.

shafts leads to a short lower streamway and a sump at −956m. Water from the cave resurges at Mohishla (altitude 70m) which has been dived but only for 60m.

West of Bzybsky, the Arabika has recently yielded more deep caves, deepest of which is the V. Iljukhin system. The nearby Kuybyshevskaya has deep shafts, large chambers and massive boulder chokes; exploration has now reached −970m, where an open shaft awaits the next visit. Its entrance is at an altitude of 2,180m, and the resurgence is the Reproa spring only 2m above sea-level. This remarkable depth potential is

SYSTEM V. ILJUKHIN

Explored between 1981 and 1984, this pothole is currently the deepest known within the Arabika massif of the western Caucasus. It has a complex series of meanders and shafts yielding three different passage down to a depth of 400m. Only one of these has been explored further, down a staircase of shafts to meet a larger streamway which can be followed to a sump at −950m. This has now been passed to a second sump at −970m, and the system is already 5km long. With its entrance at an altitude of over 2,300m, the unexplored depth potential is still enormous, as the main Arabika resurgences are close to sea-level.

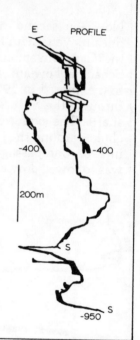

repeated elsewhere on the Arabika and Bzybsky massifs.

From Abkhasia, the southern belt of karst massifs, ranges and plateaux continue as far as the Rachinsky Khrebet, the whole being characterized by large areas of major sinks, depressions and lapies but with, so far, caves of only moderate size.

North-east of Sochi, a second major karst is formed on the Alek massif. Eight caves deeper than 200m are known to form one common hydrological system with a vertical range of 700m. The Zabludshikh and Ruchejnaya potholes converge in a system 540m deep and 2,500m long; the Ossenyaya-Nazarovskaya system now has three entrances and six streamways mostly converging at around −300m, before reaching a sump at −500m, in a cave 6,500m long. Vorontsovskaya is a 10,640m long cave on the adjacent Akhtsu massif.

The karst of the northern flanks of the Bolshoy Kavkaz starts around the headwaters of the Pshekha river, in the west, and continues in an almost unbroken belt into the centre of Dagestan. It is frequently very impressive – a fine example being the Kel'Ketchkhen sinkhole which measures 405m by 213m by 177m deep in the Cherak river basin. Deep and extensive caves have so far been found only at the western end of the range, on the Fisht-Oshten-Laganaki plateau. This contains Parjascaya Ptitsa (−535m) and Universitetskaya (2,500m long). The Dzhenty range to the north-west holds the 500m-deep Majskaya.

The Malyy Kavkaz is structurally much more varied and complicated by large-scale faulting. The karst regions are nowhere near as extensive as those of the Bolshoy and are found mainly in the Shakhadagskiu, Karabakhskaya and southern Zangezurskiu Khrebets. No major systems have yet been found, although one, Medvezaya, renowned for its dripstone, just reaches 2km in length.

Soviet Asia

It is almost impossible to conceive the sheer physical size of this territory and the amazing diversity of its landscapes. Between the Urals in the West and the Pacific Ocean in the east are 8,000km of tundra, swamp, steppe, forest, mountain and desert, and between the Arctic Ocean in the north and Afghanistan in the south the climatic conditions range from permanent frost to searing heat. Karst occurs to some extent in all these settings, much of it, until recently, very little visited by the speleologist. In many areas the potential is just suspected, while in others it is already being proved.

In the west, just south of the sprawling Urals, is the Turgay karst. It is an area of semi-arid valleys where few surface features other than sinks, springs and caves are to be seen. Further south, the adjoining areas of the Plato Ustyurt and the Mangyshlak hills provide an interesting karst of both gypsum and limestone. Many small and old caves are to be found in the broken limestone cliffs of Mangyshlak, while on the surrounding dry steppe of the Ustyurt numerous depressions, sinks, ponors and other surface features exist.

Forming a boundary with Iran are the 3,000m-high mountains of the Kopet Dag. These are mainly of volcanic origin, and the frequently mentioned caves are mostly thermal springs. The most notable of these is by the city of Bakharden: 250m long and 69m deep, it is called Bakhardenskaya and terminates in a lake whose surface temperature measures 39°C.

Rising from the centre of the desolate Peski Kysulkum are hills which were once islands in Turkestan's great inland sea. Most of these are limestone and occur as craggy outcrops surrounded now by flowing sands. Many contain small caves, shelters, depressions and minor, but welcome, springs.

The major district of all Soviet Asia is that found within the great Tien Shan mountain range. The next major range northwards is that of the Gory Altay, where the bold peaks and jagged ridges have often been likened to Switzerland. Many regions of karst are known, with the principal caves being the 2,500m long and 240m deep Altayskaya, and Geophizicheskaya (−130m, 4,200m long) and Khashim-Ojyk (6,100m long).

The second major Asian region is that encompassed within the eastern ranges of the Sayan Khrebet, Salair Krjazh, Kuznest Alatau and the adjoining hills of the Pri-Baykalye. In the former, the greatest areas of karst are situated in the often massive limestones of the Yenisey river basin. Shafts, sinks, resurgences and caves are numerous, and of the latter, many are renowned for their beautiful formations. The premier cave of the region is Kubinskaya, (−274m, 3km long). At its very western extremity, on Kuznetskiy Alatau, a recent cave discovery is the Jashchik Pandorry, with a length of 10km. Elsewhere in the range occur several of the world's longest caves in conglomerate, such as the 18km-long maze of Oreshnaya, and the 6km-long Badzejskaya. The Pri-Baykalye, to the north-east of the Sayan ranges, have areas of both gypsum and limestone karst. The main region is within the headwaters of the Lena and Angara rivers, where caves are common and are known up to 2,650m in length (Argarakan-skaya).

A vast, though little known, region of carbonate rocks lies to the north of Lake Baykal and is roughly bounded by the

Angara, Chara and Vilyuy rivers. It is a region of considerable potential but difficult of access, with extensive and unusual karst occurring in the permafrost. No cave systems of any great length have yet been recorded.

Other regions of karst in Siberia are known among the gentle hills of the Yenisekskiy Khrebet (sinks, dolines, poljes, blind valleys and depressions, in gypsum, salt and limestone), in the Pol Taymyr, in the far-eastern ranges of the Verhoyanskiy Khrebet, Khrebet Kolymskiy and the Sikhote Alin, and on the island of Sakhalin (Kaskadniy Proval, −123m). The longest cave of eastern Siberia is Proshchalnaya (2,700m).

Tien Shan

The Tien Shan is a mountain region of several major ranges each several hundred kilometres in length. To the south lie those of the Gissarsky, Zeravshansky, Turkestansky, Alajsky and Fergansky; to the north are the Chatkalsky, Kuramisky, Ugamsky and others. Many of the major peaks reach 5,000m or more, most of the others exceed 3,500m. There are vast areas of both limestone and gypsum karst, but the generally inhospitable nature of the terrain, and its distance from habitation, has precluded speleological investigations until very recent times. The resulting achievements prove that the region has the potential to become one of the world's greatest caving areas.

At the south-western end of the southern Tien Shan is the Kugitangtau range. Within this is some fine karst, where many beautifully decorated systems have been explored, including Guardakskaya (11,010m), Kap-Kutan-2 and Promezhutoch-naya, now connected into a single system 35km long. To the east, on the Bajsuntau mountains of the Gissarsky range, exploration on a plateau at an altitude of 3,500m has revealed the 565m-deep Uraliskaya, numerous shafts exceeding 100m and a resurgence some 1,800m lower down. At the western end of the Zeravshansky range lies the renowned Kirktau plateau, noted for its great fields of lapies, its many shafts and the 964m-deep system of Kievskaya. All the other ranges of the south contain great karst massifs, with features ranging from vauclusian springs, canyons and caves to sinkholes, lapies and deep shafts. On the Alajsky range, by the town of Isfara, is the 5km-long Kon-i-Gut cavern with its many great chambers up to 50m in height. Nearby are the mineralogically spectacular caves of Lyakan with their extensive encrustations of barite and celestite. In the south a series of small but interesting crags, close to the city of Osh, are limestone bastions which are considered the last remnants of a once extensive tropical karst.

KIEVSKAYA

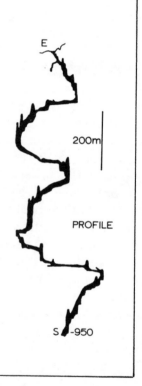

Also known as Kilsi, this is the one really deep cave of Soviet Asia, situated on the Kirktau plateau of the Tien Shan. Its single passage spirals round on itself with a sequence of long and short shafts. With its entrance at an elevation of 2,400m, it ends in a sump at a depth of 950m, still many hundreds of metres above the resurgence.

Kievskaya was explored by cavers from Kiev between 1972 and 1976. It was once thought to be over 1,000m deep, but subsequent surveys proved this to be an over-estimate. Though the Kirktau plateau is a spectacular karst, it has still not yielded any more caves even to approach the depth reached in Kievskaya.

E

200m

PROFILE

S -950

While still extensive, karst does not seem to have formed to the same degree in the northern ranges of the Tien Shan, and no really major cave systems have yet been found. The Karzhantau massif has the 260m-deep Uluchurskaya, and just to the north, on the Boroldatjau plateau, many shafts over 100m have been found.

To the south of the Tien Shan are the Pamirs. Volcanic and metamorphic rocks predominate but there are areas of both gypsum and limestone in which karst features have been noted. Few cave investigations have been made to date, but in one gypsum cave a height of +126m from the entrance has been reached.

UNITED ARAB EMIRATES

The northern extremity of the western Hajar mountains extend into the eastern coastal region of these former Trucial States. This range consists principally of dark Jurassic and Cretaceous limestones, upon which no karst features have yet been reported.

UNITED KINGDOM

See Great Britain and also Ireland, Northern.

UNITED STATES OF AMERICA

The USA has over twenty principal karst regions that contain sizeably long or deep caves. Approximately fifteen per cent of the continental United States, exclusive of Alaska, has limestone, dolomite, marble or gypsum at or near the surface. Topographically the country may be divided into Coastal Plain, Appalachian Highlands, Interior Plateaux, Interior Plains and Western Mountains and Plateaux (plus Alaska). Caves of these regions differ morphologically and reflect contrasting geological, topographic and climatic controls on their development. Because of these variations and the large expanse of the country, there is great speleological diversity not only in the characteristics of the caves themselves but in

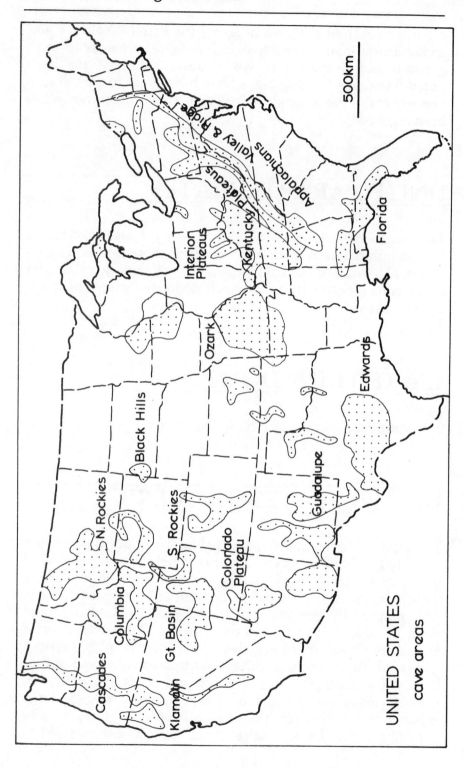

UNITED STATES
cave areas

the potential for their exploration and study.

It is difficult to estimate the total number of the caves in the US. Estimates place the total at greater than thirty thousand. In some key states where inventories are up to date, such as West Virginia, Kentucky, Missouri and Texas, the number of caves in each exceeds two thousand. Of course, most of caves are less than 1km long and less than 200m deep; however, nationally over 320 solutionally formed caves exceed 3km in length. The United States stands out as a world leader in long caves, with thirty-one exceeding 20km in length. Presently forty-nine caves have been explored to a depth of at least 150m, and seven of these exceed 300m. By far the greatest number of caves are found along the Appalachian Highlands extending from Vermont to Alabama and within the Interior Plateaux between Missouri, Kentucky and Alabama. With the exception of the Edwards Plateau of Texas, the numbers of caves are substantially less in the western half of the country.

Early cave exploration focused on caves of New England, the Allegheny Plateau of New York, Virginia and Kentucky, starting in the late eighteenth century. Scientific study of caves commenced shortly after and reached a critical plateau in the 1930s and 1940s with publication of classic papers on the origin of limestone caves by Davis, Bretz and Swinnerton. Organized speleology was born in the US with the founding of the National Speleological Society in 1941. Since then nearly all potential cave regions have been at least reconnoitred, and many have been extensively studied, mapped and documented. Yet there is still considerable potential for new discoveries in remote or less frequented areas such as Alaska and the mountainous west.

The four karst regions that have accounted for most speleological activity to date are the Valley and Ridge, and Appalachian Plateau provinces, and the Interior and Ozark Plateaux. Nearly all the longest caves are located in these classic areas; however, recent work in the Black Hills, Edwards Plateau and western states indicates that other extensive areas are just beginning to yield interesting, important and spectacular discoveries. Among these are the alpine karst of western Montana and western Wyoming, the Guadalupe Mountains of New Mexico and Texas, and the mountains of California, Idaho, Utah and Colorado.

Valley and Ridge

The Valley and Ridge karst region extends nearly the entire length of the Appalachian foldbelt. Palaeozoic limestone, dolomite and marble crop out throughout this belt, making it one of the largest areas of karst in the US and of great speleological importance.

Caves of the northern third of the region are generally small, and those in the marble of western New England and the limestone of eastern New York are relatively widely dispersed. Bashful Lady Cave in Connecticut (300m long), Eldon French Cave in Massachusetts (240m) and Morris Cave (490m) and Weybridge Cave (400m) in Vermont are the longest in their respective states. The foldbelt through New York, New Jersey, Pennsylvania and Maryland has several moderately sized caves developed in Lower Palaeozoic carbonates. Surface karst is somewhat subdued in the northern Appalachian foldbelt; dolines are small and occur in isolated groups. Small sinking streams are typically associated with caves that open from dolines, and underground drainage routes are generally short. In New York and New England, caves and drainage have been significantly affected by Pleistocene glaciation.

The degree of karst development and the character of the caves change drastically as the foldbelt is followed into Virginia and West Virginia. Here the relief reaches 1,500m, with parallel ridges and narrow intervening valleys. Limestones are up to 300m thick. The folded sequence of interbedded limestone and non-carbonate rocks has led to development of geologically simple caves, but patterns of former and present drainage through the caves may be complex. Folding has produced caves parallel to Valley and Ridge axes, along bedrock strike. Typically, master conduits are horizontal and in many caves parallel passages are of similar elevation. Most caves consist of a few relatively straight large trunk passages with dendritic networks of smaller tributaries.

In rocks of steep dip, the caves are generally long but laterally confined while in areas of gentler dip cave systems are of substantial length. Fifteen of the seventy-one American caves over 10km in length are in the Valley and Ridge region, and nine are in Greenbrier country, West Virginia.

FRIAR'S HOLE SYSTEM

This is the longest of the many very fine streamway caves in the Valley and Ridge province of West Virginia. The Friar's Hole system now extends to over 68km of passage with a vertical range of 180m. Only the top of the limestone is exposed, down the floor of a long broken valley, and many sinks provide feeders for the underlying cave. The trunk streamways lie in the Friar's Hole and Rubber Chicken sections, and both end in sumps. There are many large fossil passages which provide crossovers between streamways, and there is relatively little vertical development.

Friar's Hole and the Snedegar's-Cruikshank sections were known for many years before, in 1976, Rubber Chicken Cave was found and linked to both the older sections. Canadian Hole and Toothpick Cave were explored soon after and connected into the main system. Fresh discoveries are still being made, and the scope for extensions both upstream and downstream is considerable.

The longest caves in the region are the recently explored Friar's Hole Cave System and the Organ Cave System (59.8km, −183m). Each of these systems is a massive cave complex, Organ having two major dendritic systems within its own hydrogeologic setting. This is also true for Culverson Creek Cave System (33.5km), Windymouth Cave (23.3km) and Bone-Norman Cave System (22.7km) of Greenbrier County. However, six other long caves of Greenbrier County, the Hole (36.8km), McClung Cave System (26.4km), Benedicts' Cave (23.9km), Maxwellton Cave (15.5km), Ludington's Cave (8,900m) and Wades Cave (6,100m), form a clustered system of hydrogeologically related caves that have as yet defied connection through exploration. These have come to be known as the 'Contact Caves' because they have developed along the base of the Greenbrier Limestone where it comes in contact with the underlying shale. Other long caves of the Valley and Ridge region include the synclinal maze of the Butler-Sinking Creek Cave System (30.6km), nearby Breathing Cave (7,200m), Fallen Rock Cave (10.3km) and Perkins Cave (16.1km) of Virginia, Cuyler Cave (13.1km) of Tennessee, and Simmons-Mingo-My Cave System (10.8km, 207m deep) and Cassell Cave System (10.7km) of West Virginia. Significant new discoveries are added each year, and the potential for original exploration remains high.

Appalachian Plateau

The Appalachian Plateau karst region is parallel to and just west of the Valley and Ridge region, extending from Pennsylvania to Alabama. The Plateau is maturely dissected and has narrow, sinuous valleys incised into its surface, with local relief of 200m or more. It grades west into the Interior Plains karst province. Rock sequences include Carboniferous limestones, mostly with gentle regional dips and only shallow folds. Caves and surface karst are found along broad outcrops of the limestone. Shallow dips have promoted extensive horizontal development in many caves.

Owing to progressive incision of surface streams into the Appalachian Plateau, caves commonly have two or more distinct levels, vertically separated by up to 150m. Interconnections are provided by deep vertical shafts or multiple pits.

This is particularly true in the southernmost part of the region, where the many deep shafts have given the Tennessee-Alabama-Georgia (TAG) region the reputation of a vertical caver's paradise.

Major caves of the Appalachian Plateau include the Xanadu system (35.5km), the Mountain Eye System (25.1km) and Big Bone Cave (15.5km) in Tennessee, Ellison's Cave with its famous deep shafts in Georgia, and Fern Cave and Engle Double Cave in Alabama, each 160m deep and with over 25km of passage in the Fern System. Explorations in the TAG area have been spectacular over the last few decades, and there appears to be plenty of potential for yet further discoveries.

ELLISON'S CAVE SYSTEM

Lying in the heart of the Appalachian TAG region, Ellison's Cave is a classic system even though it is only 16.9km long and 332m deep. Its fame derives from its two spectacular shafts at either end of a through trip right beneath Georgia's Pigeon Mountain. The old Ellison's Cave was extended in 1968 by Richard Schreiber and Della McGuffin, when they discovered the Fantastic Pit – a free-fall shaft of 155m. At its foot they found a maze of streamways and large dry breakdown passage. These they explored until they found Incredible Pit from below. In 1969 they found the second entrance with a route into the top of Incredible, which proved to be 134m deep, again a beautiful cylindrical shaft. Other extensions have followed, including a high-level way into Fantastic which offers a free drop of 178m.

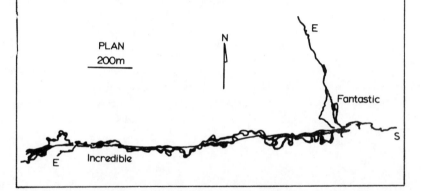

PLAN
200m

N

E

Fantastic

S

E   Incredible

Interior Low Plateaux

Perhaps the most famous of all American cave regions is in the low plateaux of Indiana, Kentucky and Tennessee. The Interior Low Plateaux region, which also includes parts of Illinois, Ohio, Alabama and Mississippi, is the hub of the vast eastern US cave belt extending from the Ozark Plateaux of Missouri to the full length of the Appalachian foldbelt. Nearly forty per cent of the long caves, each with over 5km of passage, in the United States are within the Interior Low Plateaux.

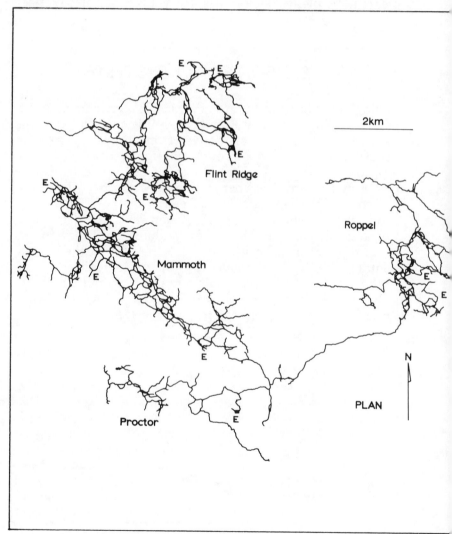

2km

Flint Ridge

Roppel

Mammoth

N

Proctor

PLAN

The region can be subdivided into three geologic units. The Mississippian Plateau, underlain by Carboniferous carbonates, begins in southern Illinois, swings through western Kentucky and Tennessee and extends into Indians. The Bluegrass Region occupies eastern Kentucky and is underlain by early Palaeozoic rock. To the south lies the Nashville Basin with Ordovician limestone, ringed by the Highland Rim with Carboniferous limestone.

Caves of the western Mississippian Plateau are numerous but generally short. There are a few long caves in western Kentucky, including Lisanby Cave (11.5km long) and Cool

MAMMOTH CAVE SYSTEM

By far the longest cave in the world, the Mammoth system now has over 530km of interconnected, mapped passages. It lies beneath the north edge of Kentucky's sinkhole plain, where almost horizontal limestone extends in low ridges to the south bank of the Green River. The system is a multiple-level complex of very low-gradient passages. The main high levels are only in the ridges and are truncated by the valleys, but the low-level streamways pass right beneath the main valleys; even so, total vertical range of the system is little over 100m. Narrow meanders alternate and intersect with dry phreatic tunnels, 10m in diameter, and massive vadose canyons. Fine cylindrical shafts connect the levels, and the low-level rivers are long and sluggish. Calcite decorations are sparse, but some of the gypsum deposits are beautiful.

Mammoth Cave has long lain open, and 12km had been mapped by 1835. Major extensions were found each year, and then in 1972 a streamway beneath the Houchins Valley made the connection with the Flint Ridge caves. In 1979 the link was found beneath the Doyel Valley to Proctor Cave lying under the long Joppa Ridge. This contained the remarkable Logsdon River, a continuous streamway 13km long, which was followed upstream to connect to Roppel Cave, beneath the Toohey Ridge, in 1983.

Further discoveries are being made regularly, and connections with a number of nearby caves are only a matter of time. Fisher Ridge Cave System (57.5km) lies just north of Roppel, and James Cave (16.5km), Lee Cave (12.2km) and Whigpistle Cave (32.5km) lie just south and west of Proctor.

Springs Cave (5,300m), but further east the famous Central Kentucky Karst has much longer caves. These include the sprawling Mammoth Cave System and its various satellite caves. Crump Spring Cave (17.3km), Hicks Cave (also known as the Hidden River complex, 31.4km long) and Grady's Cave System (19.4km) lie in that order along the south bank of the Green River just upstream of Mammoth; all are similar, massive complexes of active and fossil passages with long, low-gradient trunk streamways.

Caves of northern Tennessee and northern Kentucky are generally smaller, though Thornhill Cave is now 27km long and Big Bat Cave is just over 20km long. Further north, into Indiana, the limestone plateau once again contains exceptional caves. The longest of these include Blue Springs Cave (32.3km, with its long and spacious river passages), the comparable Binkleys Cave System (30.7km) and Sullivan Cave (15.5km), and the decorated Wyandotte Cave (8,600m).

Another area of similarly long caves lies east of the central Kentucky Karst. Several are among the longest in the US, including the Sloans Valley Cave System (39.6km), Cave Creek Cave System (24.2km), Goochland-Poplar Cave Complex (18.0km), Coral Cave System (17.0km), Wells Cave (16.3km) and Pine Hill Cave (7,900m).

Most caves within the Bluegrass region are relatively short, with few exceeding 2km in length. Some of the longest caves in Ohio lie just north of this region, but these are less than 1km in length.

The Highland Rim and Nashville Basin contain most of the long caves of Tennessee. Among these are the rambling Cumberland Caverns (44.4km) with some sections of large trunk passage, the Maria Angela Grotto (15.9km), Snail Shell Cave System (14.6km), Rice Cave (13.5km), Wolf River Cave (12.7km), Dunbar-Roy Woodard System (12.2km) and Zarathustra Cave (11.3km).

Caves of all sections of the Interior Low Plateaux share some characteristics and occur in comparable geological settings. They typically consist of dendritic networks composed of one or more large trunk passages fed by tributaries, representing evolution of mature drainage systems during development. Surface basins drained through the caves are large in area, and multiple levels in most long caves reflect progressive downcutting of surface streams. Large aggregate passage lengths for these caves are due to the well-integrated flow systems on several levels. Systematic exploration of the Kentucky caves is really only in its infancy, and many more major discoveries can be anticipated.

Ozark Plateaux

The large area of karst and caves that covers the southern two-thirds of Missouri and also reaches into Arkansas, Illinois and Oklahoma is the western extension of a continuous band of karst from the Appalachians and the Interior Low Plateaux. The region is known as the Ozark Plateaux, uplands that have become deeply dissected by surface drainage. Exposed carbonate rocks that contain caves are generally Ordovician and Carboniferous in age and are exposed both in the uplands and along major valley walls.

Caves are generally well documented in Missouri, and around 4,600 are known. Eight caves exceed 10km in length. Of these, the four longest are clustered in Perry County, Missouri: Crevice Cave (45.4km), Moore Cave System (27.4km), Mystery Cave System (25.5km) and Rimstone River Cave System (22.6km). Some of these caves are separated by only a few hundred metres, yet concerted efforts toward connecting them have so far been frustrated. They are generally dendritic in character and are developed as horizontal systems with active streams.

The longest caves in Illinois, Foglepole Cave (14.5km), Illinois Caverns (9,100m) and Kraeger Cave System (6,200m), and in Arkansas, Fitton Cave (13.1km, notable for some of its gypsum formations), Blanchard Springs Cave (11.3km), Rowland Cave (6,100m) and Ennis Cave (6,000m), are also in this region and consist of nearly horizontal, well-integrated conduits. Other long caves in the Ozark Plateaux include Carroll Cave (19.2km) in Missouri, and Oklahoma's second longest cave, Duncan Field System (7,100m), in the south-western part of the Ozarks.

Edwards Plateau

The greater Edwards Plateau of central Texas is an upland surface underlain by a thick sequence of Cretaceous and Palaeozoic limestones. With an area of 82,500km, it is one of the largest unbroken karst regions of the United States. Much of the landscape of this region consists of moderately to deeply dissected tableland, particularly along the southern and eastern plateau near the Balcones Fault escarpment. Some areas have been reduced to residual ridges, knobs and undulating hills with benched slopes.

Caves are of two main types. Those in the uplands are typically fossil, with just a few interconnected rooms and short passages and pits. In many caves, chambers are profusely decorated with calcite, as in the famous Caverns of Sonora (2,200m). The second group of caves are near present base level; many consist of lengthy conduits that formed as well-integrated drainage systems. Most of the long caves in the region, including the Powell-Neel Cave System (20.4km), Honey Creek Cave (13.3km) and Indian Creek Cave (5,500m) contain active streamways; dry caves include Inner Space Cavern (4,600m), the Airman's Cave crawl complex (3,400m) and Longhorn Cavern (3,000m). Most long caves of the Edwards Plateau are formed along one or two horizontal levels and are dendritic, with tributary passages converging onto master conduits; zones of roof collapse have blocked water flow and caused development of extensive mazes in Powell-Neel and some other systems. The deepest cave on the plateau is Sorcerer's Cave (−170m), and few others exceed 100m in depth.

Much of the plateau is roadless, and access is not easy. With its enormous area of exposed limestone, its potential for new cave discoveries is therefore very high.

Black Hills

The Black Hills, mainly in South Dakota, have a central core of igneous rocks surrounded by bands of sedimentary rocks including a Carboniferous limestone. This has all the known caves of the area, most of which are short, with two major exceptions. The longer of the two is Jewel Cave, currently fourth longest in the world. Fifty kilometres to the south-east lies Wind Cave. This has 67.8km of passage with a vertical range of 214m. Like Jewel, it is a multi-level, joint-controlled maze, and it is particularly well known for its boxworks of calcite veins projecting from passage walls. Exploration in both caves, and elsewhere in the hills, is continuing at a steady pace.

JEWEL CAVE

The longer of the two maze caves in the Black Hills of South Dakota, Jewel Cave has over 114km of passage already mapped, with numerous open leads for future exploration. It covers a vertical range of 134m in the gently dipping limestone. The passages form an incredibly complex multi-level maze, much but not all of it showing strong joint control. It is clearly phreatic and probably of artesian origin. There are many large chambers and rifts, but also many long and constricted crawlways. It has few stalagmite deposits but is famous for its spectacular displays of calcite crystals and its gypsum formations.

Most of the cave was explored by Herb and Jan Conn between 1959 and 1984. The potential for new discoveries is still enormous, and the limestone is continuous to the very similar Wind Cave nearly 50km away. Jewel Cave is now a national monument, with artificial entrances to a show cave route in the middle section of the maze.

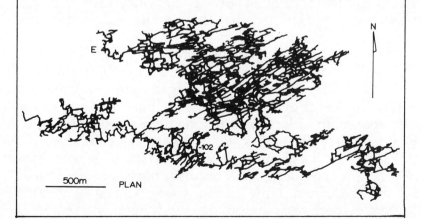

500m   PLAN

Rocky Mountains

One of the most exciting areas of caving in the western US during the last decade has been the northern Rocky Mountains. Many of the deepest caves in the country have only recently been found. The major cave-bearing rock is the 250m-thick Madison Limestone of Carboniferous age, occurring on most of the mountain ranges of Montana, Idaho and Wyoming.

Recently discovered Columbine Crawl (3,700m) in the Teton Mountains, Wyoming, is presently the deepest US cave at −429m. The karst of Silvertip Peak, Montana, has yielded a complex system of associated caves that if ever connected may become the nation's deepest. Currently the Silvertip Cave System (5,100m, −313m) is the major system. Sunray Cave (−245m) and Meander Belt Cave (−156m) are nearby. Other notable caves in the Montana Rockies include the Green Fork Falls Cave-Kathy's Icebox System (6,200m long) and Lost Creek Siphon Cave (−223m). Across in Idaho, Papoose Cave is 252m deep.

The Big Horn-Horsethief Cave System, which straddles the Wyoming-Montana state line in the Bighorn Mountains, is currently the second longest cave in the western United States (16.3km). It is maze-like and has developed largely on a single level in an Ordovician limestone. Nearby, in the same range, is Great Expectations Cave, the second deepest cave in the US at −417m.

The northern Rocky Mountains are somewhat remote, and access is often difficult owing to wilderness conditions. However, it is expected that this region will produce many more sizeable caves.

The southern Rocky Mountain region encompasses areas in Colorado, Utah and Idaho. The landscape, like that in the northern Rocky Mountains, is rugged, and caves are found within the flanks of mountains where limestone and dolomite are exposed. In Colorado, Groaning Cave (9,800m long) in the White River Plateau is the longest, and Spanish Cave (−230m), in the Sangre de Cristo Range, is the deepest. Several long and deep caves occur in north-eastern Utah. Big Brush Creek and Little Brush Creek Caves (9,800m and 7,500m respectively) are among the longest in the southern Rockies, and Big Brush is 262m deep.

Western Mountains

Outside the Rockies there are many more karst areas, of either mountains or plateaux, in the West, and caves are widely distributed through the area. Each karst region has its own characteristics and cave types, determined largely by geology.

The Guadalupe Mountains are famous for containing Carlsbad Caverns and a number of other large caves. They are developed in massive Permian limestone and typically consist of large chambers interconnected by short passageways. These caves generally have considerable vertical extent, with Carlsbad Caverns reaching −313m. New Cave (4,100m), also within Carlsbad Caverns National Park, is a smaller yet similar cave. These and many other caves in the Guadalupes are noted for their numerous massive stalactite and stalagmite deposits. Some caves in the Guadalupe Mountains are developed more horizontally and confined to thinner limestones; these include Fort Stanton Cave (10.45km) and Virgin Cave (−220m).

To the east of the Guadalupe escarpment lies the extensive Gypsum Plain of New Mexico and Texas, with many of the longest American caves in gypsum. Park Ranch Cave (3,200m) is currently the longest but recent explorations have yielded several others of similar length. It is still relatively easy to find new caves in the canyons of the Guadalupe Mountains and on the gently rolling gypsum plain.

The Basin and Range terrain, largely of faulted mountain blocks and broad valleys, occupies the Great Basin with scattered karst and caves through Nevada and Utah. The highest density of caves is around Salt Lake City but none is over 3km in length; Timpanogas Cave is noted for its beautiful displays of helictites, and the uninspiring Neff's Canyon Cave (−357m) once held the American depth record. Out in Nevada, Lehman Caves are the best known and contain an impressive number of calcite cave shields.

Further south, the Colorado Plateau is formed largely of sandstone and is deeply dissected, by the Grand Canyon among others. Beds of massive Palaeozoic limestone contain scattered mainly horizontal caves, of which the longest are Falls Cave (4,900m) and Roaring Springs Cave (3,600m) which opens into the north wall of the Grand Canyon.

South of the Colorado Plateau, another section of Basin and Range contains Edgewood Caverns (4,900m) in New Mexico.

CARLSBAD CAVERNS

The famous cave system of Carlsbad lies in the slopes of the Guadalupe Mountains in the dry country of New Mexico. Though Carlsbad is famous for its large chambers, it now extends to 33.2km of passages, reaching a depth of 316m. The system consists of a network of great tunnels and chambers; along with the connecting smaller rifts, this is a huge warren which is really a larger-scale version of the phreatic spongework found in many of the passage walls. The unusual morphology of the cave is at least partly related to its original development by migrating, natural sulphuric acid. Almost the whole cave is now dry and profusely decorated with massive calcite formations.

The entrance chambers were mined for bat guano in the first twenty-five years of the century, and at some time during that period the lower chambers and the famous Big Room were first entered. Carlsbad is now a national park, and the show-cave route goes down the massive main corridor and loops round the Big Room at a depth of around 200m. Once claimed as the world's largest cave chamber, the Big Room can no longer hold that record, though it is still extremely spectacular.

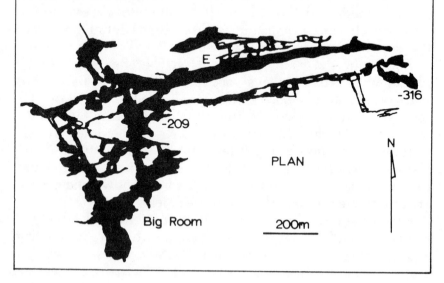

E

-316

-209

PLAN

N

Big Room   200m

There are also caves scattered through Arizona, and many of these are noted for their exceptionally beautiful calcite decorations; lying unprotected on open land, these are unusually fragile caves, and locations and details are not widely available.

Close to the Pacific coast, the Klamath Mountains contain zones of folded marble. On the California side, the fine streamways of Big Foot and Meatgrinder Caves are connected into a system 16.1km long and 367m deep. Across the state line, Oregon Caves have 4,900 of passage in a complex multi-level maze.

The Sierra Nevada of California is essentially granite but has marble in many of its marginal zones. At the southern end, Lilburn Cave is a linear system 12km long with rooms and long passages on several levels.

Western Volcanic Caves

The volcanic mountains of the Cascades Range extend from Washington down into northern California and contain large areas of basalt even though some of the volcanoes are more explosive. The finest of the lava caves are on Mt Adams and Mt St Helens, and the latter has Ape Cave (3,900m, −215m) as its longest and deepest. Most caves are single tubes, though Dynamited Cave is a multi-level system. More tubes occur in the Mt Shasta and Lava Beds area of California. Many of the caves contain ice, and Mt Rainier also has the extensive Paradise System of glacier caves.

Further east, the Columbia Plateau is a huge province of horizontal basalts in southern Indiana. There are many lava caves, notably at Craters of the Moon. Between there and the Snake River, an unusual series of volcanic fissures contain both the Crystal Ice Caves and the South Grotto, which is over 200m deep.

Other Karsts of Eastern USA

The coastal plain has large areas of low-lying Tertiary limestone, most notably in southern Georgia and northern Florida. Longer dry caves include Climax Cave (12km) and Warren's Cave (5,700m), but more famous are the many submerged caves explored by divers in the last decade. Advances in techniques have allowed exploration of the flooded conduits which connect many of the large karst springs. Peacock Springs System (6,500m) and Blue Springs Cave (5,600m) are the longest of a number of totally flooded caves.

The central lowlands have numerous small caves scattered though generally thin bands of limestone, with the largest concentration occurring west of Lake Michigan. Mystery Cave, Minnesota, is the exception to the host of short caves, as it has 19km of passage in joint-controlled mazes. In Iowa, Coldwater Cave has some exceptionally fine active streamway within its total length of 12.6km. Further south-west, Permian gypsum on the Texas-Oklahoma border contains the River Styx Cave System (2,100m) in Texas and the Jester Cave System (9,300km) across the state line. Oklahoma also has Wild Woman Cave (4,300m) in limestone in the southern part of the state.

New York state has caves outside the Appalachians in various limestones in its centre and north. The Helderberg Plateau has yielded recent discoveries in McFail's Cave (8,000m) and Skull Cave (7,200m) as well as in the old Howe's Cave System, now 4,200m long. Close to the St Lawrence River, shallow grid-pattern mazes have formed in thin Ordovician limestones, where the longest cave is the Glen Park Labyrinth System (4,300m), though five other mazes are over a kilometre long. Such lengths are not reached yet in the dolomite of the Niagara escarpment, though many shorter caves are known.

URUGUAY

Of no speleological significance.

VENEZUELA

Among all the South American countries, Venezuela is the most important for both karst and caves. Serious cave studies date back to 1952, culminating in the formation of a still-expanding national organization in 1967. Karst is extensive, in both the northern outcrops of limestone and also the extensive sandstone of the south. Of the former, major regions are to be found in the states of Falcon, Monagas, Lana, Miranda and Zulia, while lesser areas exist in Anzoategui, Merida, Aragua, Trujilo and Caraboba, and on some of the offshore islands.

The Serranía de San Luís in Falcon is the richest region yet discovered, containing many hundreds of cave systems. The area is one of dense tropical jungle surrounded by semi-arid scrubland. Dolines, uvalas, towers, pinnacles, caves and resurgences are all common. Major caves include the

305m-deep Sima del Guarataro with its 185m entrance shaft, the complex Sima Sabana Grande (−288m, 1,870m long) and the interesting Cueva de San Lorenzo (−237m, 440m long), whose entrance is in limestone but whose inner passages are in a mixture of sandstone, shale and siltstone. Just to the south, but still north of the Río Tacuyo, a more recently investigated area has yielded La Taza de la Quebrada del Toro (−120m, 1,602m long). A similar kind of karst has developed in the north of Monagas, where the country's longest cave, the 10,200m Cueva del Guacharo, is situated. Many deep caves are also known there, such as Cueva los Gonzales (−240m) and Sima del Chorro (−220m).

Marble is the main carbonate rock of Miranda in which many beautiful and quite extensive caves have developed, such as the 4,290m Cueva Alfredo Jahn, while in the limestone of the northern Sierra de Perija, in Zulia, three recently explored systems exceed one kilometre in length.

Bolivar and the northern half of Amazonas are dominated by the Roraima Sandstone, which, under the wet tropical conditions, has weathered into some of the world's most unusual karst. The whole area is dense forest, sparsely populated and difficult of access, but through the past decade four plateaux, Sarisarinama, Auyantepuy, Guaquinima and Sipapo, and the Sierra Pacaraima have all yielded quite amazing discoveries. The Sarisarinama plateau extends over an area of 35km by 25km, drained by the headwaters of the Río Cauro. Some enormous shafts have developed in the sandstone; the Sima Mayor de Sarisarinama is 350m in diameter and 314 deep, descended by a pitch of 252m, while the Sima Menor is 170m across and 248m deep with nearly a kilometre of large passage leading off from its base. The Platforma de la Sima Aonda (Auyantepuy) is perhaps even more spectacular; its surface is broken by huge rifts (of which the Sima Aonda is 362m deep), all draining to a resurgence 400m below.

The potential for future cave exploration is considerable, in both the sandstone and the limestone.

VIETNAM

Very extensive regions of spectacular karst are to be found along the northern border with China, from Ha Giang eastwards to the Gulf of Tonking and then south almost as far as the Nam Sam river. Tower and cone karst predominates with associated sinkholes, dry valleys, resurgences and innumerable short caves. Several areas of pinnacled 'karst forest' also occur. The steep-sided and often thickly vegetated towers reach 200m in height, and most have flat tops deeply etched by lapies. The towers extend into the sea and have been eroded at their bases to form even more impressive columns.

Minor and isolated areas of karst extend down the Laos border, particularly in the vicinity of the Khammouane plateau, but little is to be found south of latitude 17 degrees.

Serious but little publicized studies of the karst by the Vietnamese have been in progress for almost two decades.

WINDWARD ISLANDS

Of all the islands in this short, predominantly volcanic chain, only Barbados and Martinique are known to contain caves. Barbados is composed of coral limestone and has a well-developed tropical surface karst together with many small caves. Martinique, which is volcanic, has several short caves and some very fine coast erosion.

YEMEN

South of latitude 16, this mountainous country is formed mainly of limestone and sandstone, with some summit caps of basalt. While karst features probably are developed, even in the inhospitable climate, no caves have yet been recorded.

YUGOSLAVIA

The karst of Yugoslavia covers an amazing thirty-three per cent (50,000 sq km) of its total land area and includes the Dinaric mountains, Europe's most extensive continuous karst. Smaller, but also important karst areas occur to the east, between Nis and the River Danube, and in the Julian Alps, north of Ljubljana. Lesser regions occur in Macedonia and other parts of Serbia. Around thirteen thousand caves are known, of which five thousand occur in the northern province of Slovenia.

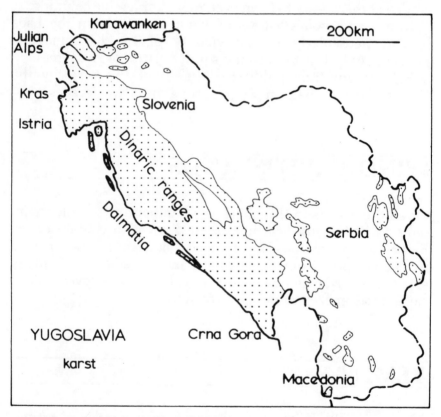

In the north-west, the Julian Alps, parts of the Karawanken, and the Pohorje are composed of limestone and are structural and scenical continuations of the mighty Austrian and Italian karst. A high precipitation has ensured a well-developed karst, particularly throughout the Julian Alps and in the Kamniške Planine region of the Karawanken. Lapies, sinks, shafts, dolines, dry valleys, depressions and caves occur in profusion, and while the country's deepest known caves are already here, there is a possible potential for systems up to 1,200m deep in the Nanos plateau. The current deepest caves are the shaft system of Brežno pri Gamsovi Glaviči (Viševnik,) −876m, 11km long) and the very complex through cave of Poloska Jama (Tolmin, −707m). No major systems are known in the Pohorje, nor in the nearby hills which extend to the north of Zagreb, but many smaller caves have yielded important palaeontological and archaeological material.

The large karst in eastern Serbia consists of much lower and more rounded hills than the Alps. Rainfall is less, but karst features occur in great diversity, as well as many caves. Notable among these are Bogavinska Pečina (Boljevač, 5,000m long) and Cerjanska Pečina (Niš, 4,000m long).

Almost all the other limestone areas do contain caves, but none yet found is of major proportions. However, not all regions, particularly in Macedonia, have been fully investigated.

Speleological history dates back to the mid 1600s, and research continued steadily over the next couple of centuries. As early as 1889 the great shaft of Kačna Jama was descended to −180m. In 1930 the world's first underground laboratory was opened in Postojna, and ever since an intensive investigation of the karst has been undertaken, with emphasis on the practical implications of living on karst.

Dinaric Mountains

This great Yugoslavian range stretches from the Italian border in the west to Albania in the south-east, a total of some 650km. These mountains then continue through Albania and down into central Greece. Physically they are divided into three − Dalmatia and the Istria peninsula, the central and high karst and an inland zone. In Dalmatia, a staircase of narrow poljes and intervening bare rocky ridges lead down to the sea from the high karst. Caves and other karst features are found throughout its length, although major caves are so far known only in the bare Velebit Planina (Ponor na Bunjevču, −534m), and near Rijeka (Klanski Ponor, −320m). A particular feature of this belt is the many great submarine resurgences, such as

SKOČJANSKE JAMA

The Skočjan cave, lying midway between Trieste and Postojna, is neither long nor deep, yet it ranks as one of the world's great karst features. The River Reka sinks beneath a cliff into a cave which continues through two giant dolines ringed by cliffs 150m high. It then enters the main river passage, which must be one of the largest underground canyons in the world – 100m deep and approaching that in width. A decorated high level then leaves the river canyon, which continues to a sump, first reached in 1890. Total passage length is 5,000m and the depth is 225m. Lighting now makes it possible to appreciate the vast size of the canyon – and Skočjan is clearly the finest show cave in Europe.

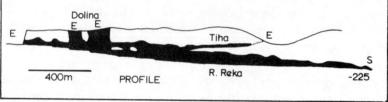

those by Split and Donja Breta, and around the islands of Brač and Korčula. At the northern end, the Istria peninsula has several fine caves within a gentler lowland karst; Raspora Jama reaches a depth of 361m.

The karst and high karst are among the wetter regions of Europe, yet the limestone is dry, barren and inhospitable, due to the efficiency of the underground drainage. The limestone ridges and karst plateaux have fine karren fields with shafts, sinks and dolines, and the poljes in between are mainly drained into large ponors and caves. At the northern end, the Kras – the classical karst – is a vast doline field hiding mainly short caves some of which are very well decorated. Just to its east, Slovenia contains the finest of the caves in a region of broad poljes and sinking rivers. Postojna Jama is the best-known cave, now a major tourist attraction, with its 14,600m of large, well-decorated, active and fossil passages. Nearby, Krizna Jama is even more beautiful and its 8,160m of passages contain twenty-two long lakes; Rakov Skočjan is a polje with natural bridges and more fine caves; Predjama has a castle in the cave mouth; and Skočjanske Jama is the finest cave of all.

Further south, in Croatia, the Jama pod Kamenitim Vratima reaches −520m with its two great shafts of 236m and 220m. Down in Herzegovina, the great poljes of Glamocko, Livanjsko and Popovo drain underground, and Vjetrenića is a 7,500m-long decorated fossil cave near Popovo. The limestone mountains of Crna Gora have the greatest potential for future exploitation; near the coast, the Orjen plateau has local relief of 1,800m, but the deepest known cave is Duboki Do (−350m). Further inland, the great limestone mountain of Durmitor has at last started to reveal its secrets: recent explorations have yielded a number of shafts, and the Jama u Vjetrena brda has been descended to a sump at −920m.

The eastern, inland zone of the Dinaric karst has lower relief with a number of large poljes, but the caves known so far are of only limited extent.

ZAIRE

Considerable areas of limestone are known in the southern half of this tropical country, but only one karst, in the Kwilu River basin near the Angolan border has been seriously investigated. Here, among numerous cones, cockpits and dolines, seventeen caves have been explored with a total of 7km of passages (Grotte sèche de Ndimba, 1,660m; Grotte de Kieza, 1,100m). On Mount Hoyo, south of Irumu, the cave system of Tsebahu Kabamba is over 1km long, and close by is the large stream cave of Yolhafiri. Karst has also been reported on the carbonates of Katanga close to Likasi, and many unexplored lava caves are known to occur on the Virunga volcanic chain.

ZAMBIA

Limestone and caves have been reported not far from Mumbwa but no other regions of karst are known to exist.

ZANZIBAR

Extensive fossil karst is known in the hills to the east of Zanzibar town, consisting of sinks, dolines and many small caves. Other small caves are known among the many coastal coral cliffs.

ZIMBABWE

Only one area of karst limestone is known, situated in a roughly east-west band to the north of Sinoia. Many dolines, depressions and small caves are known (Kua-Kua, Highlands, Ulster, Coal Hole etc), as well as the impressive Sinoia Caves themselves. This system consists of several entrances leading down to the 70m-diameter Silent Pool at a level of −68m. Divers have gained another 104m of depth without finding the floor of the pool.

1 The longest, deepest and largest caves in the world (1985)

The Longest Caves

		m	
1	Mammoth Cave System	530,000	USA, Kentucky
2	Optimisticheskaya	153,000	USSR
3	Hölloch	133,000	Switzerland
4	Jewel Cave	114,300	USA, South Dakota
5	Ozernaya	107,300	USSR
6	Ojo Guarena	85,000	Spain
7	Zolushka	80,000	USSR
8	Siebenhengstehöhlensystem	80,000	Switzerland
9	Reseau de la Coume d'Hyouernèdo	75,000	France
10	Friar's Hole System	68,070	USA, West Virginia
11	Wind Cave	67,800	USA, South Dakota
12	Organ Cave System	59,840	USA, West Virginia
13	Fisher Ridge Cave System	57,500	USA, Kentucky
14	Reseau de la Dent de Crolles	53,800	France
15	Ease Gill Cave System	52,400	Great Britain
16	Mamo Kananda	51,820	Papua New Guinea
17	Reseau de l'Alpe	51,770	France
18	Gua Air Jernih	51,600	Sarawak
19	Sistema Purificación	51,170	Mexico
20	Reseau de la Pierre Saint-Martin	50,000	France

The Deepest Caves

		m	
1	Reseau Jean Bernard	1,535	France
2	Snezhnaya-Mezhonnogo	1,370	USSR
3	Reseau de la Pierre Saint-Martin	1,342	France
4	Puerta de Illamina	1,338	Spain
5	Sima del Trave	1,256	Spain
6	Sistema Huautla	1,252	Mexico
7	Reseau Berger	1,248	France
8	Schwersystem	1,219	Austria
9	Complesso Fighiera Corchia	1,208	Italy
10	Jubiläumschacht	1,173	Austria
11	Dachsteinmammuthöhle	1,173	Austria
12	Sima 56	1,169	Spain
13	Pozu del Xitu	1,139	Spain
14	Sistema Badalona	1,105	Spain
15	Schneeloch	1,101	Austria
16	Sima GESM	1,098	Spain
17	Nita Nanta	1,079	Mexico
18	Jägerbrunntrogsystem	1,078	Austria
19	Gouffre Mirolda	1,030	France
20	Torca Urriello	1,022	Spain

The Largest Cave Chambers

	Length m	Width m	Height m	Area 002m²	Volume million m³
Sarawak Chamber Lubang Nasib Bagus, Sarawak	700	200–430	70–120	170	15.0
Majlis Al Jinn Oman	310	220	120	58	4.0
Salle de la Verna Pierre Saint Martin, France	250	260	90–140	48	3.4
GEV Chamber Torca del Carlista, Spain	250	210	20–80	51	2.9
Belize Chamber Actun Tunkul, Belize	350	100–200	60–70	43	2.5
Deer Cave (southern end) Gua Payau, Sarawak	550	100–170	100–120	70	5.0
Sótano de la Cuesta Mexico	300	90	160	24	3.0
Big Room Carlsbad Caverns, USA	550	40–130	30–80	30	1.8

Based on length/width ratios, only the first five of this list can really count as chambers, and perhaps even the famous Carlsbad Room is best described as a large passage. The Koripobi Chamber, on Bougainville Island, Papua New Guinea, may also belong in this list, but a calculation of its volume awaits an accurate published survey.

A number of giant sinkholes (notably in Mexico, Venezuela and New Britain) have volumes larger than all these chambers except Sarawak Chamber, but as they lack roofs it is irrelevant to include them in this listing.

Length, width and height are in metres, area is in thousands of square metres, and volume is in millions of cubic metres.

2 National Caving Organizations

This list is inevitably restricted to active caving organizations which have established international contacts.

ARGENTINA Organización Argentina de Investigaciones Espeleologicas, Casilla de Correo 128, Suc 1, Buenos Aires

AUSTRALIA Australian Speleological Federation, Box 388, Broadway, NWS 2007

AUSTRIA Verband Osterreichischer Höhlenforscher, Obere Donaustrasse 99/7/1/3, 1020 Vienna

BALEARIC ISLES Comite Balear d'Espeleologia, Pere d'Alcantara Penya 13 1°, Ciutat de Mallorca, Spain

BELGIUM Fédération Spéléologique de Belgique, 126 rue Royale Sainte-Marie, 1030 Bruxelles

BRAZIL Sociedad Brasiliere de Espeleologia, rua 24 de Maio 62-CJ465, Caixa Pastal 7820, São Paulo

BULGARIA Bulgarian Federation of Speleology, Bulgarian Tourists Union, 18 Tolbukhin Boulevard, 1000 Sofia

CANADA Alberta Speleological Society, 101-10531 85 Ave, Edmonton, Alberta T6E 2K5

CHINA Institute of Karst Geology, Guida Road, Guilin, Guangxi

CUBA Comite Cubano Espeleologia, Academia Ciencias Cuba Istituto de Geografia, calle 11 no 514 E/Dye, Vedada, La Habana

CZECHOSLOVAKIA Ceska Speleologicka Spolecnost, Valdstejnske Nam 1, Praha 1

FRANCE Fédération Française de Spéléologie, 130 rue Saint-Maur, 75011 Paris

GERMANY, EAST Höhlenforschergruppe Dresden, Roland Winkelhofer, Lehmannstrasse 19, 8036 Dresden

GERMANY, WEST Verband der Deutschen Höhlen und Karstforscher EV, Eugen Müller strasse 21, 4400 Munster

GREAT BRITAIN British Cave Research Association, Bridgwater TA7 0LQ

GREECE Greek Speleological Society, 11 rue Mantzarou, Athens 135

HUNGARY Magyar Karszt es Barlangkutato, Anker Köz 1-3, 1060 Budapest

INDONESIA Federation of Indonesian Speleologic Activities, Box 55, Bogor, Java

ISRAEL Israel Cave Research Centre, Ofra, DN Harei, Jerusalem 90906

ITALY Societá Speleologica Italiana, via dei Fiordalisi 6/3, 20146 Milano

JAMAICA Jamaican Caving Club, c/o Dept. of Geology, University of the West Indies, Mona, Kingston

JAPAN Japanese Speleological Society, Akiyoshi-dai Museum of Natural History, Shuho-Cho, Mine-Gun, Yamaguchi

KENYA Cave Exploration Group of East Africa, PO Box 47583, Nairobi
KOREA, SOUTH Korean Speleological Society, Dept. of Geography, Kon Kuk University, Seoul
LEBANON Spéléo Club du Liban, c/o Dr A. Malek, Ministère des Postes, Beirut
LIECHTENSTEIN Verain für Speleologie Höhlenfreunde Liechtenstein EV, Siebert Stockbrunga, Ludwigstrasse 51, 7414 Liechtenstein
LUXEMBOURG Groupe Spéléologique Luxembourgeois, Centrale des Auberges de Jeunesse, 18a Place d'Armes, Luxembourg
MEXICO Asociacion Mexicana de Espeleologia, General Cano 10A, Mexico City 18 DF. Association for Mexican Cave Studies, Box 7672, Austin, Texas 78713, USA
NAMIBIA Verain für Höhlenforschung, PO Box 67, Windhoek 9100
NETHERLANDS Speleo Nederland, Vaillantlaan 185, Den Haag
NEW ZEALAND New Zealand Speleological Society, Box 18, Waitomo, North Island
NORWAY Norske Grotteforbund, Box 321, Sentrum, Oslo 1
PAPUA NEW GUINEA PNG Exploration Group, Box 1824, Port Moresby
POLAND Kras i Speleologica, Uniwersytetu Slaskiego, ulica Miezezarskiego 58, 41-200 Sosnowiec
PORTUGAL Sociedade Portuguesa de Espeleologia, rue Saraiva de Carvalho 233, Lisboa 3
ROMANIA Institut de Speleologia Emile Racovitza, strada Mihail Moxa 9, 78109 Bucaresti
SARDINIA Gruppo Speleologico Pio, via Sanjust 11, 09100 Cagliari, Italy
SOUTH AFRICA South African Speleological Association, PO Box 6166, Johannesburg 2000
SPAIN Federacion Espanola de Espeleologia, Francesco Cambo 14, 9°6, Barcelona 3
SWEDEN Sveriges Speleolog Forbund, Box 4547, 10265 Stockholm
SWITZERLAND Société Suisse de Speléologie, Jean Claude Lalou, Ch. de Champ Manon 27, 1233 Berne
TURKEY Turkey Magara Arastirma, PK 255, Osmanbey, Istanbul
UNITED STATES National Speleological Society, Cave Avenue, Huntsville, Alabama 35810
VENEZUELA Sociedad Venezolana de Espeleologia, Apartado 47334, Caracas 1041A
YUGOSLAVIA Academie Slovenie de Science, Novi Trg 4, 61001 Ljubljana

3   Further Reading

Though the gazetteer is largely the result of a literature survey, no attempt has been made to document all the references; such would hardly have made good reading and would have doubled the text length with bibliographic detail. Most references would be available in a very comprehensive specialist library, and would then be discovered anyway by browsing under the regional classifications. It has not even proved practical to cite a key reference for each country, as for most countries there is none, while for some there would need to be a dozen quoted.

Certain publications are, however, worthy of note. For most western countries, the important cave and regional descriptions can at least be traced through the journal of the national caving organization, and the same journals carry at least reference to many major foreign explorations. The outstanding publication, invaluable to any researcher, is *Spelunca*, the quarterly magazine of the Fédération Française de Spéléologie, particularly the new series which started in 1981. *Caving International Magazine* contained some excellent review articles during its sadly short life from 1978 to 1982. And the widest English-language news reviews have appeared in the new series, since 1981, of *Caves and Caving*, published quarterly by the British Cave Research Association.

Among books, *Karst: Important Karst Regions of the Northern Hemisphere* by M. Herak and V.T. Stringfield (Elsevier, 1971) contains very academic reviews, biased away from caves, of karst in thirteen countries. In 1977 Claude Chabert published his *Les grands cavités mondiales* as a sixty-four-page *Spelunca* special supplement (number 2); this is a spectacularly comprehensive referenced set of national lists of long and deep caves. Chabert then complemented this with *Les grand cavités en roches pseudo-karstiques ou non-karstiques*, world lists of non-limestone caves, published in *Spelunca* (1980, number 3, pages 109-15). Equally valuable, with much more detail but just of the very deep caves, is Paul Courbon's *Atlas des grands gouffres du monde* (Laffitte of Marseilles, 1979). Chabert and Courbon are combining their talents to produce a new volume of world cave statistics, hopefully timed for the International Speleological Congress in 1986, and this will probably become the starting-point for any future literature search.

4 Cave Map Credits

All the cave maps have been specially drawn for this atlas and in many cases have been adapted from multiple sources, but due acknowledgement is afforded to the original surveyors, the principal of whom are as follows:

Anou Boussouil:	Collignon
Growling Swallet:	Tasmanian CC
Schwersystem:	SC Marseilles
Lamprechtsofen:	LH Salzburg
Jubiläumsschacht:	APA Varsovie
Jägerbrunntrogsystem:	Katowicki KS
Castleguard Cave:	McMaster UCCC
Reseau Jean Bernard:	G Vulcain
Reseau des Aiguilles:	SCA Gap
Reseau de la Dent de Crolles:	C Tritons
Gouffre Berger:	SGCA Française
Reseau de la Pierre Saint-Martin:	ARSIP
Gouffre Touya de Liet:	CDS Ardèche
Reseau de la Coume d'Hyouernèdo:	GS Provence
Ease Gill Cave System:	BCRA
Llangattwg Cave Systems:	BCRA
Baradla Barlang:	Aggtelek Barlang
Ghar Parau:	BCRA
Complesso di Piaggia Bella:	GS Piedmontese
Spluga della Preta:	GSCA Italiano
Abisso Michele Gortani:	US Bolognese
Grotta di Monte Cucco:	GS Perugia
Complesso Fighiera Farolfi Corchia:	GS Bolognese
Faouar Dara:	SC Liban
Sistema Purificación:	AMCS
Sótano de las Golondrinas:	AMCS
Sistema Huautla:	AMCS
Kef Toghobeit:	SC Blois
Nettlebed Cave:	New Zealand SS
Mamo Kananda:	Australian SF
Nare:	F Française S

Jaskinia Wielka Snieźna:	Koisar
Lubang Nasib Bagus:	BCRA
Sima GESM:	GES Malaga
Sistema Cueto Coventosa:	SC Dijon
Sistema Garma Ciega Cellagua:	SS Bourgogne
Ojo Guarena:	GEE Burgos
Sima 56:	Lancaster USS
Puerta de Illamina:	Maire
Hölloch:	AG Hölloch
Ozernaya:	Dublyansky
Snezhnaya:	Moscow USC
System V. Iljukhin:	Perovsky SC
Kievskaya:	Kiev ILSI
Friar's Hole System:	Melville
Ellison's Cave System:	Schreiber
Mammoth Cave System:	CRF
Jewel Cave:	Conn
Carlsbad Caverns:	CRF
Skočjanske Jama:	Hanke

Glossary

Artesian – where water moves under pressure through completely flooded cavities

Aven – shaft seen from below as a hole in a cave roof

Boulder choke – pile of inwashed boulders or collapsed roof blocks which blocks a passage

Boxwork – ribs of mineral protruding from a cave wall where the limestone has been more easily dissolved away

Calcite – calcium carbonate, the main mineral of limestones and stalagmites etc.

Canyon passage – relatively tall and narrow cave cut by free-flowing, vadose, stream or river

Cenote – flooded shaft in Central American karst

Chert – insoluble silica mineral common in lumps and bands in limestone

Cockpit – deep depression or doline in tropical cone karst

Conduit – hydrological reference to a cave as a passage for groundwater flow

Cone karst – wet tropical karst dominated by conical or hemispherical hills

Crawlway – cave passage small enough to require a man to crawl along it

Decorated cave – any cave with calcite deposits such as stalactites

Doline – closed depression in karst landscape, with internal drainage down into a cave passage or limestone fissure

Dolomite – rock similar to limestone but containing a significant amount of magnesium

Glaciokarst – karst landscape which has been glaciated during or since the Ice Ages

Gour – cave pool in which calcite is being deposited, so building up a rim or dam and making the pool deeper

Helictite – eccentric variety of calcite stalactite which grows in any direction, disregarding gravity

Histoplasmosis – disease, sometimes dangerous, caught by inhaling fungal spores in a cave containing large amounts of bat droppings

Karren – solution grooves cut into bare or solid covered limestone surface by running water

Karst – landscape distinguished by having underground drainage, commonly through caves

Lapies – bare limestone surface fretted by solution

Mogote karst – wet tropical karst with steep limestone hills, originally in Central America

Phreatic cave – cave formed below the water table, when full of water

Pinnacle karst – wet tropical karst with vertical sided blades and pinnacles of bare limestone

Pitch – vertical descent, requiring ropes or ladders, in a cave

Polje – large, flat-floored valley with underground drainage

Ponor – sinkhole which swallows the drainage in a polje

Resurgence – cave or spring where a stream flows out to daylight

Shaft – any vertical section of a cave

Shakehole – doline formed in soil overlying cavernous limestone

Shield – unusual type of stalactite in the shape of an inclined disc

Sink – cave or fissure into which water drains from the surface

Sinkhole – either a sink or a doline, depending on local useage

Spongework – extremely irregularly shaped cave formed by random solution in three dimensions

Stalactite – calcite deposit growing down from a cave roof

Stalagmite – calcite deposit growing up from a cave floor

Straw – thin, hollow, fragile stalactite like a drinking straw

Streamway – cave passage containing a stream

Sump or syphon – flooded section of cave passages completely full of water

System – a cave system is a series of interconnected cave passages

Tower karst – wet tropical karst dominated by very steep-sided limestone towers

Tube passage – rounded cave passage originally formed when full of water

Uvala – group of coalesced dolines

Vadose cave – cave formed above the water table, by a free-flowing stream

Vauclusian spring – spring where the water rises up to the surface under pressure